New Jersey Audubon Society

Administrative Office and Membership P.O. Box 125, Franklin Lakes, NJ 07417 (201) 891-1211

Directors of Conservation and Education P.O. Box 693, Bernardsville, NJ 07924 (201) 766-5787

Director of Research P.O. Box 3, Cape May Point, NJ (609) 884-2736

New Jersey at the Crossroads of Migration

Editor
Peter Dunne

Text by
Peter Dunne
Richard Kane
Paul Kerlinger

Appendixes by
Don Freiday
Patricia Sutton

Illustrated by
Diana Cann
Bill Frick
Chris Vogel

New Jersey Audubon Society
1989

Printed in the United States of America

Library of Congress Cataloging in Publication Data

New Jersey at the crossroads of migration / editor, Peter Dunne ; text by Peter Dunne, Richard Kane, Paul Kerlinger ; illustrated by Diana Cann, Bill Frick, Chris Vogel.
p. cm.
ISBN 0-9624065-0-3
1. Birds—New Jersey—Migration. 2. Birds, Protection of—New Jersey. 3. Nature conservation—New Jersey. I. Dunne, Pete, 1951— II. Kane, Richard, 1937— . III. Kerlinger, Paul.
QL698.9N48 1989
639.9'78'09749—dc20

89-14554
CIP

This book is dedicated

to the uncounted millions of migratory birds that have been lost
through nothing more than human unmindfulness

and

to the preservation of those that remain.

Contents

The New Jersey Audubon Society would like to thank the following foundations, corporations, and individuals whose generosity and concern for the habitat needs of migratory birds have enabled us to publish **New Jersey at the Crossroads of Migration.**

THE GERALDINE R. DODGE FOUNDATION

THE CHARLES ENGELHARD FOUNDATION

THE EDNA AND JACK BELASCO FOUNDATION

CHURCH & DWIGHT CO., INC.

Founding Benefactor
Howard W. Ritchie

Benefactors
Mrs. L. C. Dickinson
Marlene Lengner, M.D.

Patrons
Mrs. Howard R. Spendelow, Jr.
Mr. and Mrs. Charles F. West

Supporters
Mrs. George A. Clark
Mrs. M. L. Griswold
Heather F. and John B. Howkins
Michael J. King
MAM Foundation
North Plainfield State Bank
James P. Springer
Mr. and Mrs. John L. Strand
Mrs. Robert R. Wehner
Mr. and Mrs. Thomas M. Williams

Sponsors
David W. Babington
Mr. and Mrs. Carl Breuer
Maurette B. Hunter
Samuel C. Jones
Gail Kauffman
Dana S. Knowlton
Robert C. Martin
Newlin Smith
Ellen F. Spalt
Mrs. Donald F. Wallace
M. Stephanie Wallace
Mrs. John L. Winston

Donors
Helen Besch
Mr. and Mrs. Carl A. Bresler
Liz Claiborne Foundation
Mr. and Mrs. Charles K. Cox
Mr. and Mrs. Alfred T. Driscoll
Mr. and Mrs. John Dusenbery
Mrs. Russel J. Fosbinder
Mrs. Ralph H. Fox
Marvin R. Hyett, M.D.
International Flavors & Fragrances, Inc.
Mrs. W. G. Jens
Patricia M. La Cour
Mr. and Mrs. Charles C. Lee, Jr.
Mr. and Mrs. Milton A. Levy
Mr. and Mrs. Salim B. Lewis
Ted and Mary Marsh
Mr. and Mrs. Richard P. Massey
Lucy McMichael
Morgan Guaranty Trust Co.
Mr. and Mrs. John R. Morton
Russell W. Myers
Mr. and Mrs. Franklin E. Parker
Niranjana Patel, M.D.
Mrs. and Mrs. Jeffrey B. Savit
Deborah Shields and Robert Feldman
Muriel A. Tomkins
Mr. and Mrs. Gerardus H. Wynkoop II
Joseph Zellers

We would also like to thank several hundred additional New Jersey Audubon members and friends not listed above for their generous gifts and spirited support for Operation Flight Path, our major conservation project, and for this resulting publication.

GERALDINE R. DODGE FOUNDATION

July 14, 1989

An Invitation to the Reader:

You hold in your hands a book on birds whose song we must hear or our own song will become off-key, broken, and dissonant. The work might be called, "The Home State." For New Jersey is the staging or nesting place for 349 species of birds, 94 percent migratory, some journeying 20,000 miles a year to complete planetary appointments for eating, mating, and fledging of young.

I think this book is a litmus of how little we have learned in the quarter century since Rachel Carson published *Silent Spring*, an invitation to local vigilance. New Jersey is America's most densely populated state, even with the sparsely settled million-acre Pinelands, and we are not only a physical crossroads for fabulous bird migrations but at a temporal crossroads where we will determine whether our use of the land will allow the birds to persist . . . and ourselves with them. New Jersey is the nexus mine shaft, and we have three hundred and fifty kinds of canaries to warn us before we choke on the lethal fumes of our own combustion.

Twelve years have passed since Pete Dunne's best day, 4 October 1977. He counted 21,800 hawks of twelve species as they flew over Cape May Point. On the beach nearby a woman was sunning herself. Toward five o'clock she folded her beach chair, walked over, and asked Pete what he was doing. He replied, "Counting hawks." She said, "Have you seen any?"

We are all like the woman on the beach near the lighthouse, so near but yet so far from detecting the hidden world all about us. The hawks are not so obvious, even a southward migration of 150,000 of them, for they fly at considerable elevations over a month, and they don't form a flying wedge as do Canada geese.

In 1983 the Dodge Foundation provided funding for the Hawk Watch project and guide, a forerunner for Crossroads of Migration. Woven into the pages of this report card on how we are doing is a prescription of what we must do, knowing that we dwell on sacred ground.

Three of the commissioned cover works for the Dodge Foundation's annual report have acknowledged avian miracles in our midst that are natural wonders as grand as the big canyon in Arizona but more subtle, like many charms of the Garden State. In 1982 we featured the Great Blue Heron by Stefan Martin in a sanctuary with 100 live nests in Sussex County. In 1982 the vast October hawk migrations were suggested by the brush of Natalia Kapelyan in the flight of a juvenile female peregrine. In 1988 the northward surge of 11 species of shorebirds to sandy beaches along Delaware Bay to gorge themselves on the rich grey-green-pink eggs of horseshoe crabs was hinted at in a watercolor by Ann Williams.

This invitation to Pete Dunne and his colleagues' meticulous plea for habitats in our home state responds to René Dubos's urging that we "think globally and act locally." It is also a fitting local tribute to the thesis of *The Home Planet* which contains the words and images of cosmonauts and astronauts upon viewing our blue-green orb from outside its gravitational pull. Fred Hoyle foretold our response in 1948:

> "Once a photograph of the Earth, taken from *the outside,* is available . . . a new idea as powerful as any in history will be let loose."

Pete Dunne's book is part of the explication of the idea of the planet as a living tapestry. In the spirit of Aldo Leopold and John Muir, he has written a song which we must hear or we shall relinquish the prospect of survival with grace.

Sincerely,

Scott McVay
Executive Director

95 MADISON AVENUE • POST OFFICE BOX 1239 • MORRISTOWN, NEW JERSEY 07962-1239
TELEPHONE: 201-540-8442 • FACSIMILE: 201-540-1211

njas
new jersey audubon society
790 EWING AVENUE • FRANKLIN LAKES, N.J. 07417 • (201) 891-1211

New Jersey's reputation as a center for business, industry, advanced education, and the arts is well-established. But our state plays another pivotal role, one of global consequence. New Jersey is a strategic crossroads for millions of migrating birds.

Twice each year, hundreds of species of birds using the ancient flyways of the eastern United States make a critical stopover in New Jersey. These birds depend on New Jersey's abundance and diversity of habitats. They use our forests, marshes, grasslands, lakes, islands, river valleys, wetlands, mountains, and beaches for resting and feeding during the most dangerous and stressful period in their lives—their arduous journey between breeding and wintering grounds.

New Jersey can boast the largest hawk migration in North America and the greatest massing of shorebirds this side of Alaska. It is the state that hosts the World Series of Birding and draws visitors from around the globe whose sole purpose is observing birds.

As New Jersey residents, many of us enjoy and have come to accept the rich diversity of birds in the same way we accept and appreciate the multitudes of wildflowers that line our roads. It is one of the things that makes New Jersey special, but a thing that can no longer be taken for granted.

As the organization that has made New Jersey's natural history our business for close to a century, it is our duty to warn New Jersey citizens that something we take for granted is in jeopardy. In many parts of New Jersey once-common birds have simply disappeared. Why? Some point to the destruction of tropical rain forests. Some maintain that the integrity of nesting areas here in North America has been compromised. Very likely the growing pressures birds face during all phases of their existence are taking a toll and populations are collapsing. But it is a fact that without proper habitat at key links along the migration routes—without New Jersey—global protection efforts focused solely on breeding and wintering grounds are doomed to failure.

With the generous support of its members, the Geraldine R. Dodge Foundation, Charles Engelhard Foundation, and the Edna and Jack Belasco Foundation, New Jersey Audubon has mobilized its resources in an effort to understand the decline of our migrating birds and to offer a local corrective for a global problem. One result of this effort is this book.

We hope this publication will provide information and guidance to government and private land-use planners, developers, and conservation organizations as well as landowners, so that the habitat needs of migratory birds can be considered. To the extent that the habitat needs of migratory birds are *included* in decisions involving land use and acquisition, our project will be a success and, once again, New Jersey will have taken a leadership role in solving a major environmental problem.

Thomas J. Gilmore

Thomas J. Gilmore
Executive Director

Acknowledgements

This book was a group effort involving a host of dedicated and talented individuals. Though they gave of themselves without thought of compensation, they deserve, at the very least, the thanks and recognition of New Jersey Audubon.

We thank and acknowledge:

The many photographers who generously supplied the fine photos illustrating this booklet. This talented cadre includes: Norm Adams, Alan Brady, Barbara Brennan, Richard L. Ditch, Marion Glaspey, John Kricher, Charles Mohr, Sam Orr, Arthur Panzer, O. S. Pettingill/Vireo, Leonard Lee Rue, III, Frank Schleicher, Dave Sibley, Chris Sturm, Clay Sutton, and Dave Ward.

The several illustrators who put pen points to parchment in the name of bird protection—Bill Frick, Chris Vogel, and Diana Cann, and we are proud to display their talents here.

The General Drafting Company of Convent Station who graciously provided the map of New Jersey found on page 74.

Dr. John Kricher, ecologist from Wheaton College, Massachusetts, and N.J. Audubon Society board member, and Don Freiday, Scherman-Hoffman Sanctuary naturalist, for reviewing this manuscript for error.

Christine Camy for her considerable typesetting skills and Joan Rogers of Master Design Studio, who crafted a book from raw type and decks of photographs.

Dorothy Clair, editor, who had a hand in every stage of the development of this book and without whose help the project never would have known fruition.

And, finally, to the members, staff, and the board of the New Jersey Audubon Society, particularly Jane Brady, Jennie and Clay Gaitskill, Sally W. Lee, George J. Mullen, Jr., Patricia Morton, John D. Stewart, Martha Webster, and to all the others who supported us in this project.

LIKE A GAME OF MUSICAL CHAIRS

There was a game we used to play in grade school—perhaps it is still played. It was the teacher's safety valve for rainy day lunches; the graphite rod that dampened prevacation energy levels. It was called *musical chairs.*

This is how it worked. Two rows of chairs would be lined up back to back. The class would form a line that circled the empty seats. The teacher would put a record on the phonograph (they didn't have compact disks, yet) and play a song. When the teacher lifted the needle, the music stopped and everyone would scramble for a seat.

There was a catch of course. There weren't enough chairs to go around. Someone was always left standing, and that meant having to leave the game. Then, everyone remaining would get up again. Another chair would be removed. The music would start to play. And we'd march. It was all lots of fun. Except when you lost.

SHOREBIRDS

FRANK SCHLEICHER

□ Shorebirds! Long-distance migrants who have depended upon New Jersey's tidal wetlands for centuries as a refueling point during their pan-hemispheric flights.

GREAT EGRET

RICHARD L. DITCH

□ Great Egret—once hunted to the brink of extinction by plume hunters, now, through protection, a common resident of New Jersey's coastal marshes.

Migration is a lot like this. The seasons call the tune and the great journey begins. Along the way, there are places where birds sit down, to rest and feed. And just like musical chairs, there aren't enough places to go around. We keep removing them. Every spring and every fall some birds are left standing. They lose—and losing takes on a bit of significance when it means you die, when it means a species diminishes, when it means that the quality of life on earth drops another rung.

Loss. That is the heart of the matter. Not potential loss, not minimal loss—significant, measurable loss of birds. Only in the last few years have scientists begun to take stock of what has happened to North America's once abundant migrating birds. The picture is frightening.

OPERATION: FLIGHT PATH

□ Red-eyed Vireo—once touted as the most abundant nesting bird in the eastern woodlands—has suffered an 80% decline in Bergen County's Green Brook Sanctuary.

□ Hooded Warbler, like the Red-eyed Vireo, a once common nesting species that winters in the tropics, has disappeared as a nesting species in Green Brook Sanctuary.

RED-EYED VIREO
DAVE WARD

HOODED WARBLER
JOHN KRICHER

In Rock Creek Park, Maryland, a typical, eastern hardwood forest, the number of resident migratory birds dropped 70 percent between 1948 and 1983. An isolated instance? Unfortunately, not. During this same period in Bergen County New Jersey's Green Brook Sanctuary, Red-eyed Vireos, once touted as the most abundant nesting bird in the eastern woodlands, went from 24 pair/100 acres in 1950 to 5 pairs in 1983 (Serrao, John. 1985. "Decline of Forest Songbirds." *Records of New Jersey Birds,* Vol. 11, No. 1). Hooded Warbler, another common resident, went from 20 pairs in 1958 to zero in the late seventies. (Source: David Wilcove, Wilderness Society, personal comments). Like many other birds, both of these species nest in New Jersey and winter in the neotropics (the tropical forests of Central and South America).

Shorebirds (plovers and sandpipers), a diverse group of birds that pass through New Jersey twice each year on their journey to and from the arctic tundra, show the same precipitous loss of numbers. Sanderling, the well-known sandpiper of New Jersey's beaches (although they do not nest here), declined by an astronomical 80 percent between 1974 and 1983. Whimbrel declined by 60 percent; dowitchers, 40 percent. (Source: Dr. Brian Harrington, Manomet Bird Observatory).

WHIMBREL
FRANK SCHLEICHER

SANDERLING
FRANK SCHLEICHER

DOWITCHER
FRANK SCHLEICHER

□ Sanderling! The common, animated sandpiper of the beach declined by 80% in just ten years.

□ Whimbrel numbers have been reduced by 60%.

□ Dowitcher have decreased by 40%.

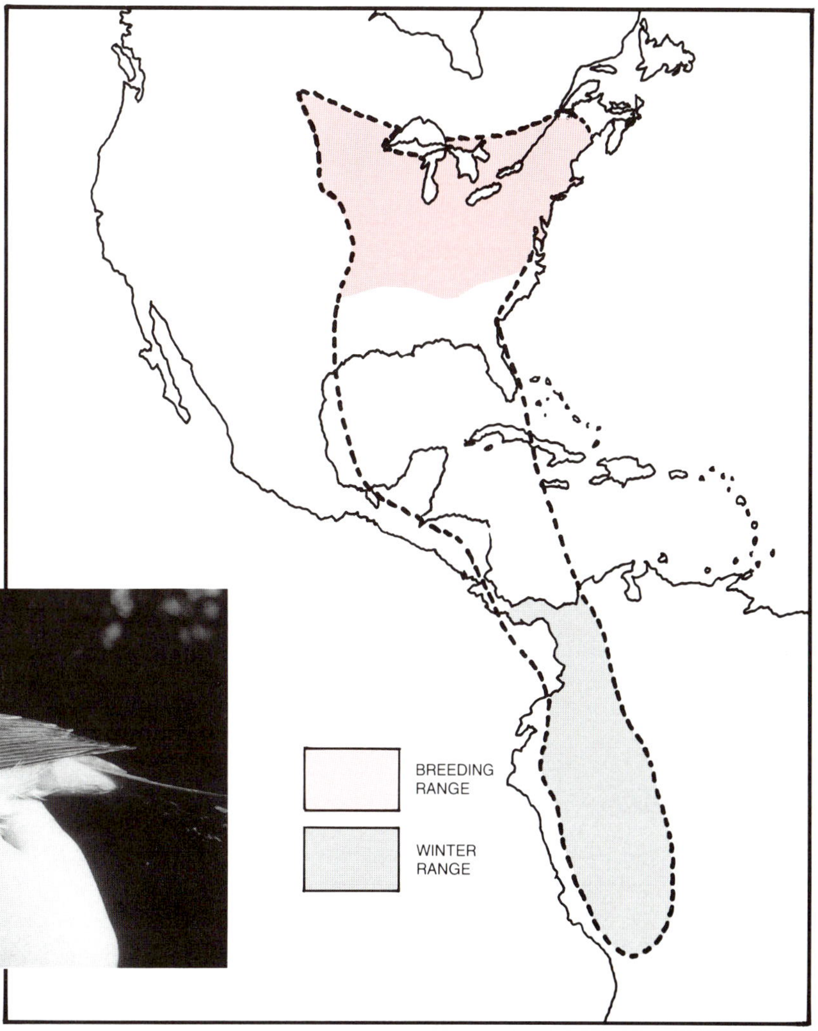

SCARLET TANAGER
DAVE WARD

□ From May to August, the Scarlet Tanager lives in New Jersey. Between November and February, it is found in the tropical forests of Central and South America.

What's happening? Simple. We're loosing ground, *habitat*—the woods and fields and marshes—that North America's birds need to survive, and the loss is everywhere they turn. Take the example of the Scarlet Tanager.

From May to August, this stunning, black and crimson bird lives in the hardwood forests of North America. Between November and February it lives in the tropical forests of Central and South America. If either the northern or southern forest is lost, the tanager cannot survive. The specializations that make the bird suited for woodland habitat make it unsuited for any other. It cannot feed on a suburban lawn, like the robin, or on a sandy beach, like a sandpiper, any more than a chemical engineer could take a job repairing diesel engines. So long as sufficient woodlands and rain forest exist, the tanager will survive. Deprive it of these, and it cannot exist.

The key to protecting birds is protecting their habitat. For many years, protection efforts focused on protecting the nesting habitat. But birds still declined. This strategy only focused on roughly one third of their needs and their lives. Eventually, private organizations and government agencies have begun efforts aimed at protecting habitat in

WOOD THRUSH
LEONARD LEE RUE III

DAVE WARD

□ The key to protecting birds is protecting habitat. Nesting habitat is important for birds like this wood thrush... so is proper habitat in the tropics (above) where it winters.

□ WHAT HAS BEEN OVERLOOKED IS THE IMPORTANCE OF HABITAT *DURING* MIGRATION. These Glossy Ibis and Egrets are feeding in the Cape May Migratory Bird Refuge owned by the Nature Conservancy.

the tropics, where many northern breeding birds spend our winter (another one third of their lives). But even this measure is insufficient.

What has been largely overlooked in our slowly developing global protection strategy is the importance of habitat *during* migration. Without it, birds will not reach their northern or southern destinations. WITHOUT IT, GLOBAL PROTECTION EFFORTS SERVE FOR NOTHING. Protect all of the breeding woodland in North America and all of the rain forest in Central and South America and tanagers will still decline *unless the habitat they need during migration is also preserved.*

New Jersey is a critical link in an incredible journey. Remove this link and we pull an irreplaceable chair out from under millions of migrating birds whose stake in the game is survival.

But I See Birds All the Time!

If you are like most New Jersey residents, you are conscious of birds. You probably saw some today. Thinking back, you might not recall that birds are any less numerous today than, say, ten years ago. For some species, this is true. In fact, populations of some species have even increased.

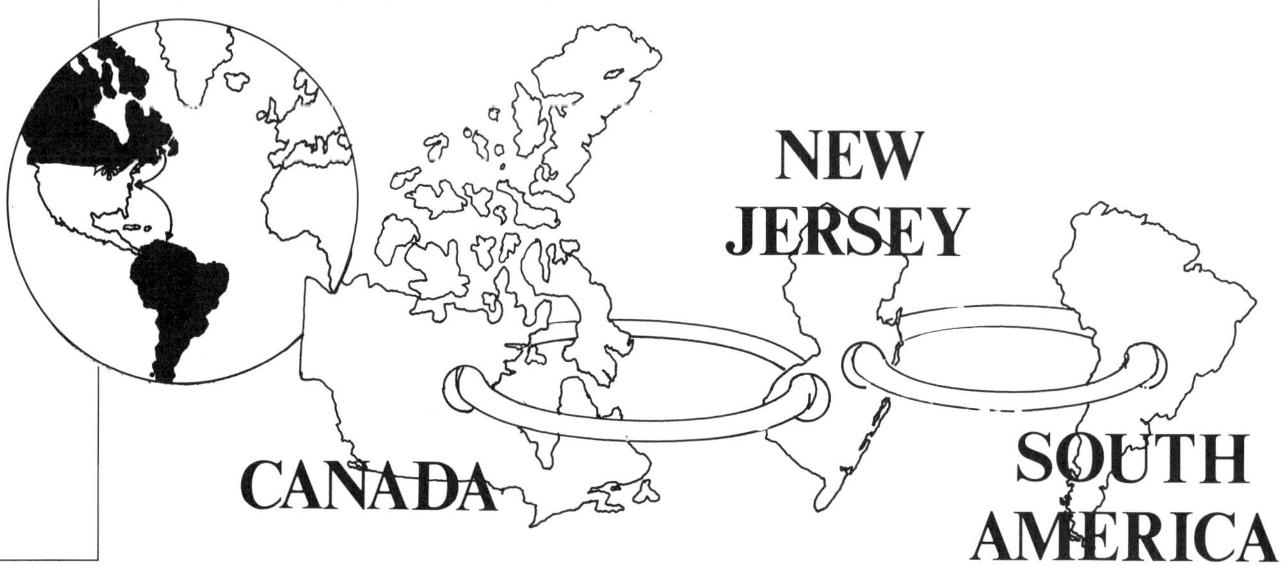

The problem is that the birds that are most commonly seen are also the species that are increasing—birds like robins, house sparrows, cardinals, mockingbirds, crows, starlings. They are increasing because as we change the habitat to suit us, we also change it to suit them.

But most birds cannot survive in a suburban environment. Birds like the Scarlet Tanager. Are we saying that a tanager is *better* than a robin or a cardinal? No, of course not. There is no value judgment involved. The concern of scientists and environmentalists is *species diversity*. Human activity, by and large, diminishes the species diversity. If twenty acres of woodland are converted into twenty one-acre suburban lots, perhaps five species of birds will be able to live there, which is fine. . . .

Except that now, sixty-five species of woodland birds cannot.

Can't the Birds Just Move Somewhere Else?

It might seem that if birds are forced to move from one place, they can simply go somewhere else—but they can't. Healthy habitats are used right up to their capacity. Nature uses her resources the way most people use their budgets—right up to the limit. There isn't much waste or much left over.

Let's go back to the twenty-acre woodlot. Say we cut down half of it. The birds from the half that was cut cannot just move into the other half any more than ten families could just move in with ten other families, if half of a twenty-unit development was destroyed in order to plant a forest.

SCARLET TANAGER

DAVE WARD

ILLUSTRATED BY DIANA CANN

□ If habitat is lost, birds, like this Scarlet Tanager, can't just move to another place—unless they displace another Scarlet Tanager.

Why Should Birds Be a Major Concern, Anyway?

Think of the earth as a room. It started as an empty room, and over millions of years, it became decorated as we see it today. The furnishings (birds, plants, trees, rivers) are not permanent. They change gradually over thousands of years. Sometimes, fixtures replace fixtures that are old and outmoded or no longer fit the grand scheme. Sometimes entirely new furnishings are added, and, occasionally, older ones are removed and just not replaced.

But nature, the grand designer, has a certain decorative style. Nature's taste is inclined toward a greater diversity and a richer world—a pleasure to our senses, a challenge to our minds, a storehouse of goods that we need to survive.

Birds are a major part of the earth's riches. They are found on every corner of the globe. They are a critical link in the interlocking web of all living things. This web is humanity's critical habitat—as important to our survival as woodland is to the tanager. Diminish the habitat, and you diminish the chances of survival.

YELLOW WARBLER FRANK SCHLEICHER

□ Think of the earth as a room. The decorations are the plants and animals that have evolved over millions of years.

ILLUSTRATED BY BILL FRICK

LAND FILL

NORM ADAMS

□ The changes we make benefit one species—our own.

TROPICAL RAINFOREST DEFORESTATION

DAVE WARD

□ Where nature adds furnishings, we remove them.

Since man, the grand manipulator, has appeared, there is no longer one master designer. There are two: Man and Nature. But our senses of design differ. The changes we make in the room tend to benefit one species—our own. Where nature adds furnishings, we remove them. In the process, we are slowly destroying the interlocking web that is our critical habitat.

When we have finally removed all of the chairs in the room, and all other creatures have left the game, man can triumphantly claim the last chair for himself . . .

RICHARD L. DITCH

□ The Bald Eagle, the national bird, and a New Jersey nesting, wintering *and* migratory species.

. . . the last chair, in an empty room.

The purpose of this book is simple. We need to help in the global effort to protect migrating birds by maintaining this critical link—New Jersey. To do this we want to impart a mindfulness of the importance of New Jersey's environment to every resident.

In separate chapters, this book will discuss:

1. MIGRATION: THE GREATEST SHOW ON EARTH—A foray into the phenomenon of migration—the great global strategy for survival that shifts millions of birds across hemispheres twice each year.
2. A GOLDEN PLOVER AT EBB TIDE—The story of the second most important day in a bird's life.
3. THE INCREDIBLE FLYING MACHINES; THE INCREDIBLE JOURNEY—An exploration of how creatures whose weight is measured in grams can undertake the incredible journey, and what they must do, and what they must have, to succeed.
4. NEW JERSEY—ONE GREAT REST STOP ON THE GLOBAL HIGHWAY—The critical nature of our small state and its historic importance to migrating birds.
5. SILENT SPRING; EMPTY AUTUMN—A grim accounting of the damage that has already been done to North America's birds and of the causes.
6. SAVING SEATS IN A CROWDED ROOM—A blueprint for protecting migrating birds to insure that they will always have a place to sit in New Jersey.

It took years to notice the decline in migratory birds. It took more years to piece together the causes of the decline. Now comes the next step. Now we have to take the measures necessary to stop it.

MIGRATION: THE GREATEST SHOW ON EARTH

DUNLIN — PETE DUNNE

□ Dunlin flock at Holgate. Dunlin are arctic breeders who winter in and migrate through New Jersey.

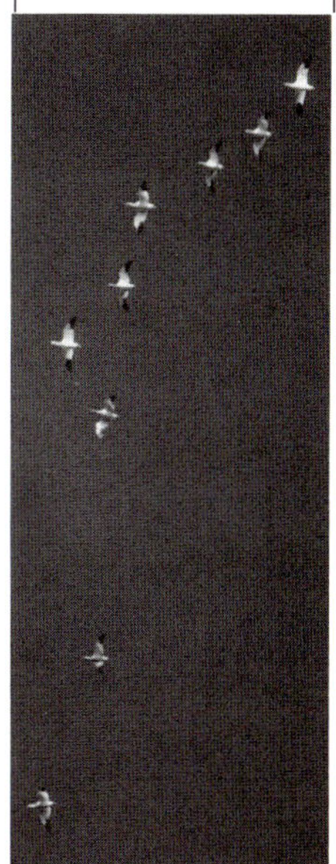

SNOW GEESE — PETE DUNNE

□ Snow Geese over Cape May Point.

Twice each year, billions of birds redistribute themselves across the earth's surface. This global strategy for survival is called **migration.** It does not happen overnight. It does not happen everywhere the same way. But it happens, and its mechanics are as predictable as the phases of the moon or the changes of seasons.

Long before the nature of migration was understood, people living in northern, temperate areas wondered at the seasonal nature of birds. They noted that birds common in summer disappeared when weather grew colder and reappeared when temperatures mellowed. People knew that many animal species follow this same pattern and

UPLAND SANDPIPER — FRANK SCHLEICHER

□ Upland Sandpipers, like this one in Somerville, New Jersey, begin their fall migration to Argentina in July.

that their disappearance was linked to hibernation—another strategy for survival. They mistakenly concluded (for lack of a better explanation) that many birds must hibernate, too.

The successor to this theory was hardly more plausible. Troubled, perhaps, because nobody had actually ever found a hibernating bird, people reevaluated their observations and postulated another theory to account for the disappearance of birds—one that was in accord with observed phenomena. They concluded that every fall, birds migrated . . . to the moon.

We now know that, with few exceptions, birds do not hibernate and, to the best of our understanding, they do not make lunar voyages. But many species *do* migrate. They vacate large geographic areas and disperse into other regions hundreds or, more commonly, thousands of miles away. New Jersey, for example, has 349 species that occur regularly in the state. Only 21 of these do not migrate. If we consider just the 201 species of birds that nest in New Jersey, 180 are migratory.

ROUGH-LEGGED HAWK
FRANK SCHLEICHER

□ Rough-legged Hawk in the Hackensack Meadows.

LESSER YELLOWLEGS
FRANK SCHLEICHER

□ The first fall migrant is usually the Lesser Yellowlegs. It reaches New Jersey on or about June 24.

FRANK SCHLEICHER

□ Hudsonian Godwits, dowitcher and "peep" (small sandpipers) take wing.

Migration! The word is alluring, charged with excitement. Every autumn, when the morning air is chill, we see the restless, feeding birds in our shrubs and gardens; we observe the rising and falling banners of geese or massed flocks of blackbirds moving like smoke on the horizon. It stirs something deep inside us, a restlessness, an eagerness for adventure and distant places.

The migratory urge is triggered by the amount of sunlight reaching the earth. It moves the hands of an internal clock set in the brains of birds (and many other creatures) and the setting differs for different species. We tend to think of migration as something that happens for a few short weeks, about the time the leaves fall off the trees or when flowers bloom in the spring. The reality is quite different.

The first fall migrants reach New Jersey just a few days after the summer solstice—June 24-25. These are shorebirds, sandpipers, far northern breeders whose migration will carry them across the equator. But the first destination of their long journey is New Jersey—where for thousands of years they have foraged in our rich tidal marshes. These early migrants are so early, in fact, that they may actually pass other shorebirds still heading north!

By July the migration of some species of shorebirds (Lesser Yellowlegs, dowitchers, Whimbrel) will peak. By August, songbird species join the southbound exodus; in September hawks and many other species join the rising tide of southbound birds, and in October and November waterfowl and seabirds hold center stage. In December, freezing conditions and passing cold fronts squeeze the last migrants out of the north. In January and February, only the wandering northern finches are left—which is fortunate. Because during the first days of February, the first northbound birds of spring arrive, vanguards of a mirror movement of birds that will continue on into July.

Spring and fall migration are protracted affairs; the onset and end of one overlap the other. What most people think of as spring and fall migration periods is really just the peak—rush hour at the Crossroads of Migration.

Where do New Jersey's migrants herald from? The most simple answer is also the most accurate: EVERYWHERE! Everywhere in North and South America except, perhaps, for the far American West. In the fall, great masses of songbirds pour out of New England and eastern Canada. Most hawks and eagles herald from the area east of the Great Lakes—but virtually the entire arctic peregrine population (Alaska to Greenland) funnels down the Atlantic Coast. In Cape May, they regularly record over 600 of these endangered falcons every fall.

A number of birds head east before starting south. Black Terns and the western subspecies of the Willet, a shorebird, are regular fall migrants, and irregular migrants are common and expected every year.

□ By February, the first northbound birds of spring arrive. The Turkey Vulture is one of them.

TURKEY VULTURE

FRANK SCHLEICHER

PEREGRINE FALCON

FRANK SCHLEICHER

□ Virtually the entire arctic Peregrine population migrates down the Atlantic Coast every autumn. Over 600 of these federally designated endangered birds have been counted in Cape May during a single autumn.

OPERATION: FLIGHT PATH

In spring, New Jersey hosts shorebirds from as far south as Tierra Del Fuego, at the very tip of South America, and Osprey that wintered in Argentina. Warblers that spent November through February from Central America to the Caribbean pulse through our state in waves. And just offshore, thousands of Greater Shearwaters—birds whose origins are several small islands in the South Atlantic—are hugging our coast.

CAPE MAY WARBLER
DAVE WARD

□ In spring, warblers (like this Cape May Warbler—named after Cape May, New Jersey) pulse through our state in waves.

For some, the great leaps that separate their northern and southern lives are accomplished in incredible, nonstop feats—2,000 miles in

GREATER SHEARWATER
FRANK SCHLEICHER

□ In May, many thousands of Greater Shearwaters are moving north forty miles off New Jersey's coast.

one pinpoint leap of faith. For many others, increments are shorter, progress a matter of days and weeks on the road. But by whichever road, the long road or the short one, migration is taxing and energy sapping. If the food a migrating sandpiper or warbler needs is in short supply, its migration will end in failure. It may reach its destination too late to nest. It may never reach there at all. And, it may die.

Migration is a *very* protracted phenomenon, and it is accurate to say that there is not one day during the year when birds are not migrating either north or south in New Jersey. Sometimes, they do both. By spreading migration out over a long period many more birds can use the same key areas. They simply do so at different times. It's a little bit like staggering the end of a work day. If every

FRANK SCHLEICHER

□ There is not one day during the year when birds are not migrating, north or south, in New Jersey. Pictured: Hudsonian Godwit.

work place let out at 5:00 p.m., New Jersey's highways would come to a halt. Letting workers out at different fifteen-minute intervals between 3:00 and 5:00 p.m. eases congestion.

For some species, like many shorebirds, adults leave the arctic first. Hundreds of thousands reach New Jersey by late July. Rested and refueled, they depart by mid-August—just ahead of newly arriving immature birds, birds hatched a few short weeks ago.

Birds of prey—hawks, eagles, falcons, vultures—use a similar strategy by reverse roles. For several species, young birds go first; adults follow a week or two later. Still, bottlenecks do occur. On 4 October 1977, 21,800 hawks were counted in Cape May.

Hawks and, very probably, many songbirds use another strategy that reduces congestion and possible competition as birds pour through New Jersey—they use different routes. There are two main hawk migration corridors through New Jersey. The first runs along the ridges and valleys of the northwest; the second along the coast. Approximately 70 percent of the birds riding the crest of Kittatinny Ridge—New Jersey's pathway in the sky for 30,000 hawks a year—are adult. In Cape May, where an estimated 200,000 hawks a year pass each autumn, 80-90 percent of the birds are immature. [Editor's Note: Immature birds are fully grown and fully capable of flight but not yet of breeding age. Some birds reach breeding age the year after they hatch. It takes some birds, like eagles, four or five years to reach maturity.]

The overwhelming number of warblers, vireos, thrushes, and flycatchers passing through our state every fall are young birds.

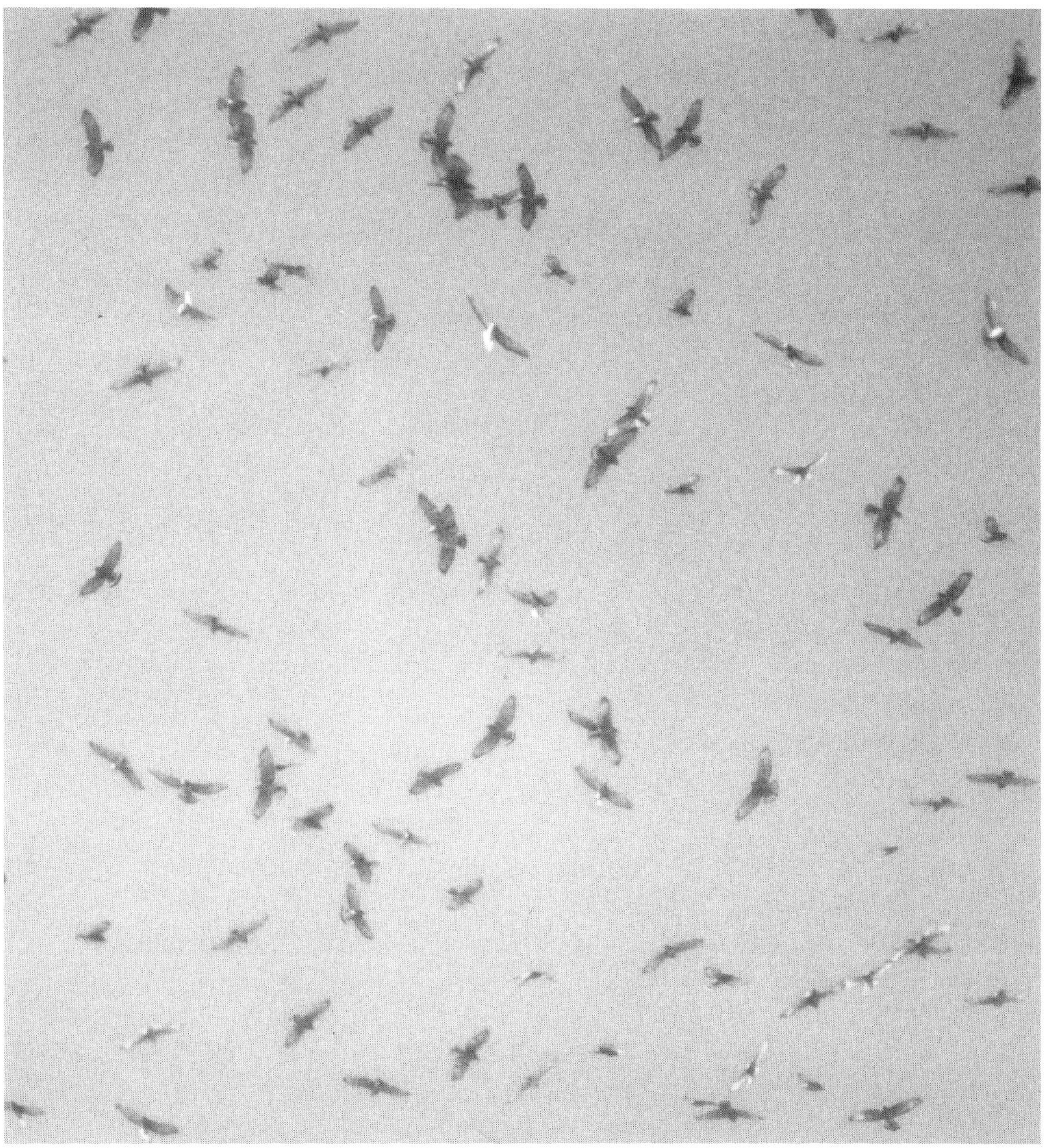

PETE DUNNE

□ Some of the 21,800 hawks counted by Cape May Bird Observatory staff at Cape May Point on 4 October 1977. Pictured: Broad-winged Hawks.

BLACK-AND-WHITE WARBLER
DAVE WARD

□ Most of the birds that migrate through New Jersey in the fall are immature.

SHARP-SHINNED HAWK
FRANK SCHLEICHER

□ Perhaps 75% of all Sharp-shinned Hawks fail to reach maturity. Pictured: immature Sharp-shinned Hawk in migration along New Jersey's Kittatinny Ridge.

This does nothing but underline the importance of New Jersey's habitat. Young birds lack experience. They are not efficient feeders. They need more time and they use more energy to secure food. They make mistakes and a second chance means the difference between death and survival. Immature birds need all the help they can get. For centuries, they have relied on New Jersey to pull them through—and it did. The proof is seen every autumn and every spring.

The various strategies have at least this in common: they work. They don't work for every bird. Migration is a perilous time for birds, the riskiest venture in their risk-filled lives. Sixty percent of most songbirds hatched in any given year fail to live one year. For birds of prey, the toll is even higher. Perhaps 75 percent of all Sharp-shinned Hawks fail to reach maturity—harsh terms by our standards. But despite the hard terms, the technique works. Species can divide their lives between widely spaced locations and link them through the air and key places to rest and feed.

At least, the strategy used to work.

A GOLDEN PLOVER AT EBB TIDE

The bird was crouched in a shallow depression just above the reach of the morning tide on a beach that linked the two towns. A casual stroller would never have seen it. Even a very attentive passer-by, one skilled at sifting sand-colored birds from sand, could not have been certain (the bird blended in so well with her surroundings). If the bird remained very still, there was nothing to distinguish her from the darker patches of sand. And the bird remained very still. This was the second most important day in her young life. This is the day she would die.

ILLUSTRATED BY CHRIS VOGEL

Her life had begun on a bluff rich with flowers that fell sharply into the Arctic Ocean. To her left, the channels of the Mackenzie River delta shimmered like a tangled web of silver. Overhead, snow buntings spiraled and the shadows of jaegers haunted the earth. But the first thing the plover had seen when she opened her eyes (before she was three minutes old) was the gray arctic sea.

A writer would find this significant: the fact that the bird would begin and end her life within sight of two oceans thousands of miles apart—but it would mean nothing to the bird. Her mind had no gift for recalling places and events outside of immediate time and need, the thing we call memory. In the orbit of her existence, the bluff and its carpet of flowers were four months behind (or six months ahead). If her fortune were different, the knowledge of the place and its significance would awaken in her when it was time—on some day in May during a spring that she would never see.

*Right now, the incredible mechanism that was her brain told her that it was time to migrate. It told her how she should do this, what route she should take, where she should rest and feed, and, ultimately, it would tell her when and where to stop. Even now, dying, the brain said "***south***"; so this was the direction she faced, her head drawn back, eyes closed, the tip of the wing that had been injured barely touching the sand.*

Migration is a time of great stress and peril for birds. The dangers are magnified by inexperience. Only a small fraction of the birds born in any given year survive to maturity, harsh terms by human standards. Migration is the first major cut in the rank of birds, and survival depends as much upon innate skills as it does upon finding places **en route** *that meet the needs of migrating birds—places to rest and feed. Unfortunately, these migratory way stations often conflict with the habitat needs of a more aggressive species, our own. Each year, because of man and his activities, migration becomes just a little bit more perilous. Fewer and fewer birds make the cut.*

But **our** *bird's injury was an accident, a near brush with an ancient enemy that claimed the life of another plover feeding nearby. Though not the intended prey, the hard leading edge of the Peregrine Falcon's wing struck the young plover's wing as the falcon cut through the panicked flock. Two of the very fine bones near the end of the wing were broken.*

The bird could still fly, but not as well as a Golden Plover should. When the restless flocks departed on the last day of August to begin their great journey south, she remained. Three weeks passed, and though not fully mended she could wait no longer. She left on a morning with the air filled with snow and migrating sea ducks.

Flying alone, without other birds to take their turn breaking trail, she used energy quickly. Friction dragged at her wings. She tried flying with the sea ducks but their speed was not her speed. Even injured she outflew them. She was a Golden Plover and, injured or not, she was one of the most incredible mechanisms of flight ever conceived by God or man.

The bird put down in a newly harvested wheat field in northern Manitoba. The place was rich in grasshoppers made torpid by the deepening cold. Two other plovers were there, juveniles like herself. One bird had a broken leg. In a week, when they left, she joined them. Although her energy reserves were insufficient, her need for company was great.

Flying as three was easier than solo flight, but the bad wing was a handicap and it sapped her energy quickly. Over northern Maine, she dropped back and pitched into a potato field. Strangely, the other plovers joined her. A bond had been formed. They were a team. But food was not plentiful. Winter was driving the worms they sought beyond the reach of their bills. Two days later, the threesome took wing again, heading south and east.

The storm that forced them to make landfall on an island off the coast of Massachusetts lasted two days. During that time, the birds huddled under a sandy lip on the lee side of a grassy knoll. They spent the week following the storm searching the short-cut grass for the food they desperately needed (avoiding as best they could grounds keepers and men who moved about in motorized carts).

Food was difficult to find. Grounds keepers wage a ferocious war against insects that might mar a fairway. When the two birds finally lifted off heading out over open water, the injured bird hesitated. Her strength was not up to theirs. Locked in her brain was the number of miles that lay ahead, and the same brain told her that her fuel, the layers of fat she had amassed, was not sufficient for such a journey. She hesitated, staring after her retreating companions.

One of the pair called back to her, a low, plaintive whistle. She answered immediately and lifted off in pursuit.

For several hours she managed to hold the hemisphere crossing pace but slowly her energy played out. Land, she knew, lay somewhere to the west. She fell behind, trailed the pair for a time, and then turned away, heading west. It took most of the night, fighting a northwest headwind before she found the land she sought.

As her strength waned, she lost altitude until her wings barely cleared the water. Twice, a wave nearly claimed her.

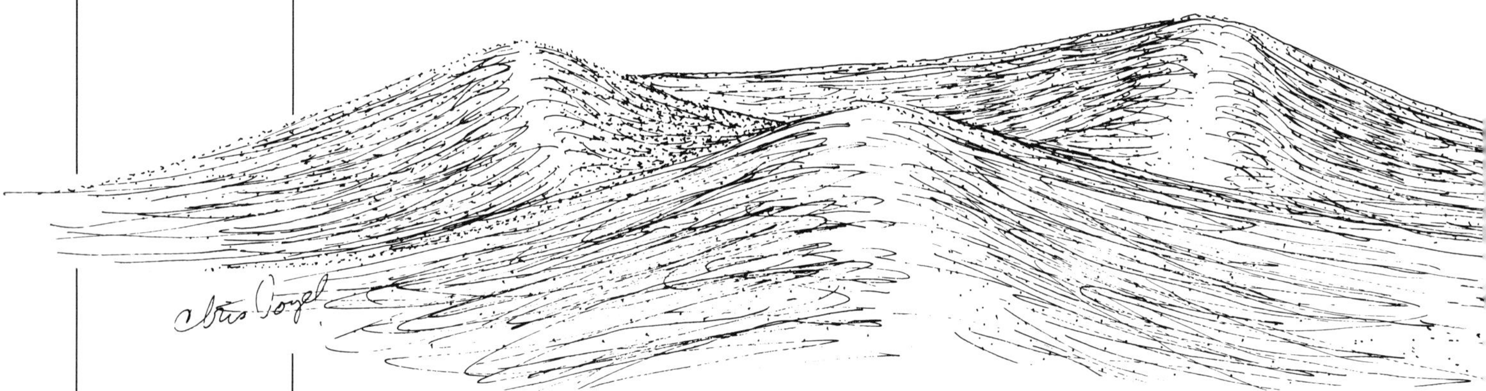

Once, she actually landed on the water. But plovers are not phalaropes. They are not made to float. She had no choice but to continue.

She reached the beach at the very tip of New Jersey just before dawn—a small bit of fortune. An hour later and the sharp-eyed gulls would have spied her, driven her into the water, and fought for her remains. The exhausted bird collapsed just above the reach of the cresting tide and was asleep immediately.

She awoke near midmorning, stood, and moved unsteadily toward the water's edge. Earlier in the season, she might have hunted more successfully, but the cold air had already driven most of the temperature sensitive worms into deeper waters. A constant stream of strollers kept the bird moving; distracting her, forcing her to use energy she didn't have just avoiding their feet.

Frustrated, the bird flew to the other side of the dunes where grassy, well-pooled meadows lay. She explored the shallow pools for insects or worms but found nothing. The meadows had been repeatedly treated to control mosquitoes leaving nothing for birds to eat. Most migrating birds stayed only a few minutes before flying off in search of more productive areas. But now flight was beyond the plover's capacity. Returning to the beach she crouched in a shallow depression above the tide and the heaviest pedestrian traffic. This is where she spent the rest of the day, and after the sun fell to earth, the night.

When morning came, she remained where she was. There was no energy left to search for the food that might have sustained her life. What was left of her life would be measured in hours.

Near noon, a young couple, enjoying the last days of Indian summer, came down to the beach accompanied by a large red setter. Though the dog was supposed to be on a leash, few people paid attention to the regulation. The couple walked within twenty feet of the bird without seeing her. Even the coursing setter missed her twice, but on its third pass, the dog nearly tripped over the huddled form.

The bird felt the dog's inquisitive nose move over her back. She tried to stand but could not. With an effort she succeeded in opening her eyes. In front of her, to the south, was the sea.

The dog ran down the beach at full gallop, eager to show its prize, nearly bowling the woman over in its excitement.

"What do you have there, Loki?" the man demanded. "Hold up, damn it! Let me see. Stop!" With incredible effort, the dog managed to immobilize its fore-end (but the hindquarters were simply out of control).

"Oh, my God!" the woman exclaimed, horrified . "It's a dead bird. A dirty old dead bird. Drop, it," she commanded, dealing Loki a sharp slap on the nose, "Drop it!"

Loki dropped it.

The morning sun spread a golden wash over the still form, making its underparts glow warmly. Even now, it was beautiful.

"Huh, I wonder what kind of bird it is?" the man mused aloud. "It's not a sea gull."

"It's a dead one," his companion said impatiently, as she stepped in front of Loki whose interest in the bird was unabated. "Come on," she demanded, "let's get away from here before he picks it up again. It probably has some kind of disease."

An hour later, a wave propelled by the incoming tide claimed the bird and carried it into Delaware Bay.

Many hundreds of miles to the south, a single golden plover landed on a shallow coral flat and proceeded to feed, clumsily because of exhaustion and because of its broken leg.

THE INCREDIBLE FLYING MACHINES; THE INCREDIBLE JOURNEY

Paul Kerlinger
NJAS Director of Research

OSPREY

FRANK SCHLEICHER

The eminent evolutionary biologist Ernst Mayr once said that migration is a "perilous" time in the life of birds. Mortality during migration, especially among birds making their first journey, often exceeds mortality in any other season. For example, mortality among Ospreys (*Pandion haliaetus*) undertaking their first autumn migration can be greater than 50 percent (Poole, A.F. and B. Agler 1987). This means that only one of two young of the year live through their first migration. Among migrants that live to migrate again, the probability of survival is greater, although they must still undertake two "perilous" flights each year.

CHARLES MOHR

Why is migration a "perilous" time for birds? Why do so many migrants die? Mortality is a result of starvation, competition, drowning, predation, disease, shooting, storms, collision with man-made objects, or combinations of these factors. After suffering injury, near starvation, or disease a bird may live, but not return to its breeding ground. This is equivalent to death, because the migrant does not realize a net reproductive gain. That is, it does not leave offspring to perpetuate its genes and those of its species. In this chapter I describe and discuss migration distance, causes of mortality, enroute behavioral ecology, and physiology of migrating birds. The importance of habitat is then discussed as it relates to these topics.

RED-TAILED HAWK

BARBARA BRENNAN

□ Gunshot Red-tailed Hawk in Somerset County.

□ Dead birds found beneath the Cape May Point lighthouse after a foggy night in 1939.

□ Migration is a perilous time for birds. Mortality in the ranks of young Osprey (far left) can exceed 50%.

WIRE-KILLED EGRET

FRANK SCHLEICHER

□ A Great Egret dangles from the guide wire of a tower it struck in the Hackensack Meadows.

How Far Do Birds Migrate (or, How Far Is Far Enough?)

The rigors of migration are proportional to the distance a bird migrates. Birds that make long-distance migrations face greater risks than short-distance migrants or birds that do not migrate. Evolution through natural selection has shaped the behavior, morphology, and physiology of migrants such that they are adapted to a bewildering array of conditions. Migration distance is no exception. Because of the rigors and dangers of migration, birds usually do not migrate farther than they have to.

The distance birds migrate between breeding and nonbreeding locations varies within and among species and species groups. By this I mean that different species migrate different distances and that different individuals within a species migrate different distances. Although a few songbirds and raptors migrate very long distances, the shorebirds and terns migrate the farthest. Within these groups there are species that do not migrate at all and there are others that migrate more than 25,000 kilometers round trip.

In general, there is no typical migration distance. Of more than 650 species of birds that breed in the United States and Canada, some 250 or more fly into the neotropics (Mexico and southward) during the winter. Birds that breed in New Jersey include long-, middle-, and short-distance migrants as well as birds that do not migrate. Among the long-distance migrants are Broad-winged Hawk, Least Tern, and Louisiana Waterthrush. All migrate south of the United States with some individuals reaching the South American mainland. A Broad-winged Hawk that moves between northern New Jersey and Colombia must fly more than 5,000 kilometers each way. To do so, it flies around the Gulf of Mexico through Texas. Other migrants cross the Gulf of Mexico thereby reducing the total

GULL-BILLED TERN

FRANK SCHLEICHER

□ Gull-billed Terns are among the world's great long-distance migrants.

distance they must fly. Each year tens of millions of warblers, thrushes, tanagers, grosbeaks, and others cross the Gulf of Mexico. Although these migrants realize shorter flights, they face rigorous water crossings of more than 1,000 kilometers.

Middle- and short-distance migrants from New Jersey include the Red-tailed Hawk, Forster's Tern, Black-crowned Night-Heron, and Brown-headed Cowbird. Few, if any, of these individuals leave the United States, and some individuals of given populations do not migrate at all. These migrants have one-way flights ranging from less than 100 kilometers to more than 2,000 kilometers. Finally, birds such as the Ruffed Grouse, Carolina Wren, Carolina Chickadee, and Northern Cardinal usually do not migrate.

Among middle- and short-distance migrants distance is variable within and among populations. For example, an examination of banding data from across North America for the Red-tailed Hawk reveals that the more northerly populations move farthest, sometimes leap-frogging over more southerly

RED-TAILED HAWK

FRANK SCHLEICHER

□ Immature Red-tailed Hawks often migrate farther south than adults.

populations that are not as migratory. In addition, immature (first-year) birds often move farther than adults. Immature birds are scarce in central New York and northern New Jersey during winter because they move to more southerly climates. Similarly, female American Kestrels migrate farther than males. This phenomenon is called differential migration. It is not known why this occurs, but some biologists believe it has something to do with social dominance, whereby subordinate immatures are forced out of an area by dominant adults. One explanation of why male kestrels do not migrate as far as females may be the males' need to remain close to nesting territories so that they can reclaim prime sites prior to the breeding season. At present, we do not know how extensive differential migration is, nor do we know why it occurs. Comparative and experimental studies may help us to better understand differential migration.

To complete migration a bird must either make a few long flights or several shorter flights. The distance birds actually travel in one flight varies greatly. For birds that migrate at night, flight-distance rarely exceeds 500 kilometers. With flights of less than 300 kilometers being the norm, most middle-distance migrants must make at least three to six stopovers during migration. Long-distance migrants may make as many as ten to fifteen stopovers, although some species make only two or three.

Radar and radiotelemetry studies (Kerlinger, P. 1989. *Flight Strategies of Migrating Hawks.* Chicago: University of Chicago Press) have shown that raptors fly between 100 and 350 kilometers per day. With perfect atmospheric conditions, they may traverse 500 kilometers. These birds migrate mostly by day because they depend on thermal updrafts. These ascending columns of warm air allow hawks, eagles, falcons, storks, pelicans, and other soaring birds to gain several hundred meters before gliding for several kilometers. During this type of flight relatively little energy is used, but the birds are constrained to fly when and where thermal currents are available.

Mortality during Migration

Populations of animals fluctuate with changes of the birth and death rate. When the two are equal a population is stable and when the death rate exceeds the birth rate a population declines. Reams of information are available on the breeding biology of birds. In particular, we know how many eggs birds lay, how many hatch, how often nesting attempts succeed, and how many young live long enough to fledge (leave the nest of their parents). We know relatively little, however, about how long most birds (or other animals)

FRANK SCHLEICHER

□ Birdbanding is a key tool for measuring migration and mortality in birds. Pictured: Common Snipe.

live or how they die. Because of this paucity of information, mathematical models of animal populations are not always realistic.

As a result of long-term population studies, banding research, and some field studies, we are now acquiring information on mortality and longevity that will promote more accurate population models. These models are important because they help wildlife biologists to formulate management plans.

Banding and visual studies have shown that mortality during migration can be great. The example given above for Ospreys is informative. Most bands from yearling Ospreys (found dead) were recovered during autumn migration between the species' breeding and wintering areas. This strongly suggests that young birds, undertaking their first migration, are particularly susceptible to accidents, disease, storms, and starvation. Other studies show similar patterns. Banders have also shown that first-year birds weigh less than adults, which further supports the idea that immature birds often starve during migration.

While studying songbird migrants, I have observed mortality resulting from drowning. The study was conducted on Horn Island sixteen kilometers into the Gulf of Mexico off Biloxi, Mississippi. Migrating songbirds must fly more than 1,000 kilometers to successfully complete the flight between southern Mexico

COMMON YELLOWTHROAT

FRANK SCHLEICHER

□ An exhausted Common Yellowthroat that landed aboard a party boat 80 miles east of Barnegat Light. Thousands (if not millions) of migrating birds drown every year.

and the southeastern United States. Many of these migrants pass over or stop on Horn Island. On several days I found dozens of songbirds of about twenty species washed up on the beach. Larger birds such as tanagers, grosbeaks, and cuckoos were identified easily because whole bodies were recovered. Only flight feathers of small birds like warblers and vireos could be found in the surf line. These birds had evidently become fatigued during transgulf migration or encountered adverse weather and simply dropped into the gulf. Exhausted migrants are observed regularly from oil-platforms in the gulf and from boats at sea. Each year thousands (if not millions) of migrants succumb during this long and arduous journey.

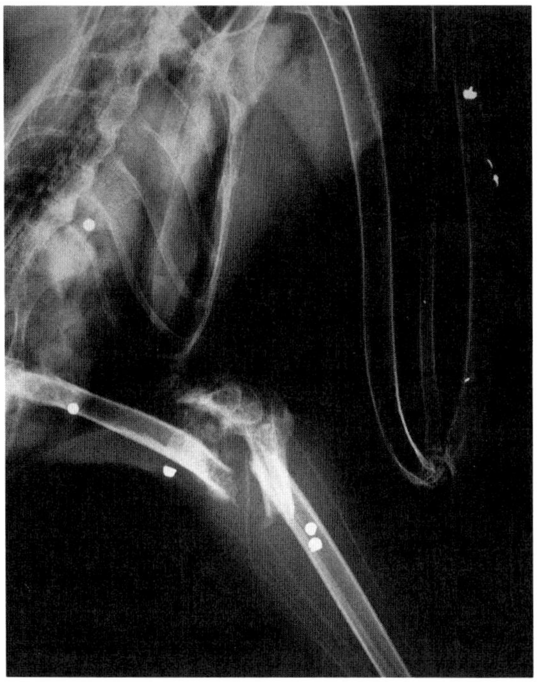

BARBARA BRENNAN

□ X-ray of a gunshot bird of prey. Lead pellets show up as bright white spots. Despite full protection, birds of prey (and other protected species) continue to be shot.

SNOWY OWL

FRANK SCHLEICHER

□ Mortality in Snowy Owls caused by collisions—mostly with man-made objects—may run as high as 50%.

Some mortality is directly attributable to human activity. Although the shooting of raptors has been illegal for several years, William S. Clark has found that a small percentage of the migrants he bands at Cape May Point, New Jersey, are shot. A wholesale slaughter of migrants still occurs in Italy, Spain, Greece, the Middle East, the Philippines, Mexico, and many other places in the world. When species like hawks pass through populated areas, they can be shot because they cannot rest and hunt in safe habitats.

I recently examined seventy-five Snowy Owl carcasses (migrating and wintering birds) from Alberta and Saskatchewan and found that about 12 percent of the birds were shot. (More than 50 percent died of injuries resulting from collisions—mostly with man-made objects.) Ten to 15 percent is the figure cited in most studies of mortality causes among raptors.

Other human-induced deaths of migrants are caused by collisions with cars, towers, windows, buildings, wires, and other structures. Tower kills have been used to study the body condition and species composition of nocturnal migrants. At towers that stand more than 300 meters above the ground, kills can be massive with single structures accounting for hundreds of birds in one evening. In southern New Jersey we are finding that low-flying migrants are impacted heavily by automotive traffic. Each winter several dozen hawks and owls are found dead and brought to the Cape May Bird Observatory. Invariably, birds like Northern Saw-whet Owls and Hermit Thrushes have been hit by cars. How this mortality is impacting the general population of these and other species is unknown.

Predation by other birds and even mammals can be important in some locations. At Cape May it is easy to witness predation. During autumn migration Sharp-shinned Hawks and other bird-eating raptors may eat thousands of songbirds during their passage through the cape. Observing these hawks from the Cape May Hawk Watch makes it obvious how much predation actually occurs. A large (but unknown) proportion of the bird-eating hawks that pass have full crops.

SAW-WHET OWL

SAM ORR

□ Northern Saw-whet Owls, a bird that flies very low, are habitually struck by cars. Twenty-five were killed in Cape May alone during November and December of 1988.

PETE DUNNE

□ The remains of one of the many thousands of songbirds killed and eaten by migrating hawks in Cape May, New Jersey.

Even predatory birds are not safe from predators. Small hawks such as American Kestrels and Sharp-shinned Hawks are sometimes taken by larger hawks and owls. At Cape May Point, New Jersey, where raptors are forced into close proximity to each other, this occurs seasonally. Most often Peregrine Falcons and Cooper's Hawks prey upon smaller Sharp-shinned Hawks and American Kestrels. Gustav Rudebeck, the well-known Swedish ornithologist, has reported that during the autumn migration through southern Sweden some individual Peregrine Falcons specialize in taking hawks such as European Sparrowhawks.

Because it is difficult to determine the magnitude of predation upon migratory (or any) birds, it is difficult to conclude how predation impacts populations of migrants.

EASTERN KINGBIRD

FRANK SCHLEICHER

□ Eastern Kingbird, a common New Jersey migrant, consumes a dragonfly (another common migrant).

Predation is probably more important in areas such as Cape May and less important elsewhere. Thus, the causes of mortality probably vary from location to location depending upon the number of predators present and other factors relating to migratory flight.

Behavior, Duration, and Fat Deposition during Stopovers (or "Fat's Where It's At")

Following a migratory flight, birds must decide whether to continue migration or stopover to rest and replenish energy reserves. This decision need not be conscious, although it must be based on the bird's energy reserves, weather, and the availability of suitable sites in which to rest and feed.

Picture a Black-and-white Warbler flying from Yucatan, Mexico, to its nest site in New Jersey. During its flight over the Gulf of Mexico, it must fly nonstop for more than fifteen hours. If a cold front with associated northerly winds impedes its migration, it could use all of its energy reserves and fall into the gulf or be delayed by several hours. Upon arriving at the southern coast of the United States, these birds are in a depleted condition.

For a bird to migrate 10,000 kilometers or even 100 kilometers, it needs energy to fuel its internal muscular "engine." The longer the migration, the more energy is required. Whereas some energy is available to a migrant as a form of sugar in the blood, this source is enough for only an hour or two of flight. The most important energy source for migrants is fat.

Fat is stored in several locations in a bird's body. The largest fat deposits are stored just under the skin (called subcutaneous fat) at the top of the breast and on the abdomen, although some is stored around internal organs such as the liver and heart. Subcutaneous fat is visible through the skin of many birds. By blowing breast or abdominal feathers aside, fat can often be seen as a creamy white or yellowish deposit. Fat birds also feel fat. They are round and plump and the sternum (breastbone) is not easy to feel and does not protrude. The sternum of a starving bird or one that has flown a long distance protrudes. When this occurs a bird has used all of its body fat and some of the protein in its flight muscles. The use of muscular protein as an energy source is a last resort prior to exhaustion or death.

A migrant such as a sixteen-gram Kentucky Warbler can fly 200 to 250 kilometers on a gram of fat. By carrying four to six grams of fat, an individual can fly between 800 and 1,500 kilometers before having to use its own muscles as fuel. The same migrant must have one to three grams of fat for a one-night flight. To cross the Gulf of Mexico, a Kentucky Warbler must have at least three to five grams of fat available. With only three grams it must also have a tail wind, or it will not complete its journey. Deposits of this magnitude represent increases in body mass of 10 to greater than 40 percent.

In 1986 I conducted a study of several warbler species in a small (four-five hectare) coastal forest in southwestern Louisiana. It is frequented by large numbers of migrants following their 1,000+ kilometer flight from Yucatan. By netting, banding, and weighing birds, we found that a small percentage (20 percent) of warblers such as Worm-eating, Hooded, Prothonotary, Kentucky, and Ovenbird stopped over for one to two days, whereas Northern Waterthrushes, Tennessee Warblers, and others almost never made stopovers. The latter species migrate farther north and rarely were recaptured. This may be because they immediately continued northward following their arrival.

Migrants that stopped over for one or more days almost always gained weight. These migrants tended to weigh less and have less fat than other individuals that continued their migration without making stopovers. (Stopover duration of some of the warblers I studied in Louisiana averaged 1-2 days.) Birds that stopped over one to two days or longer gained weight at an average of 3 percent of their body weight per day. This is equivalent to a 150-pound man gaining 7.5 pounds per day (at 5 percent gain per day). Thus, most migrants of these species will make stopovers of three to five days with regularity at several sites during autumn or spring migration.

Rufous Hummingbirds deposit fat at a similar rate to warblers. Researchers in the western United States found that these tiny migrants are capable of increasing their body mass from 3.5 to 5.0 grams in less than a week. This is an increase of more than 40 percent of body weight or an average of about 4 to 7 percent of body weight per day. During stopovers these birds defend territories, much as they do during the breeding and wintering seasons. Such defense is costly, but assures that the indivudal will have a continued source of rich and dependable nectar.

Although information is difficult to acquire for shorebirds, it is likely that they also gain around 2 to 5 percent of their body weight per day during migratory stopovers. Studies done by Brian Harrington and others from the Manomet Bird Observatory demonstrate that individual Red Knots stopover for up to two or more weeks along the Delaware Bayshore

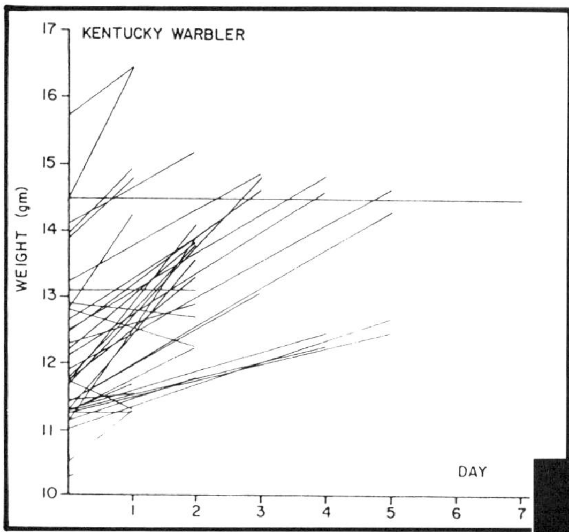

DAVE WARD

□ Figure 1: To cross the Gulf of Mexico, this Kentucky Warbler must have 3 to 5 grams of body fat to burn as fuel.

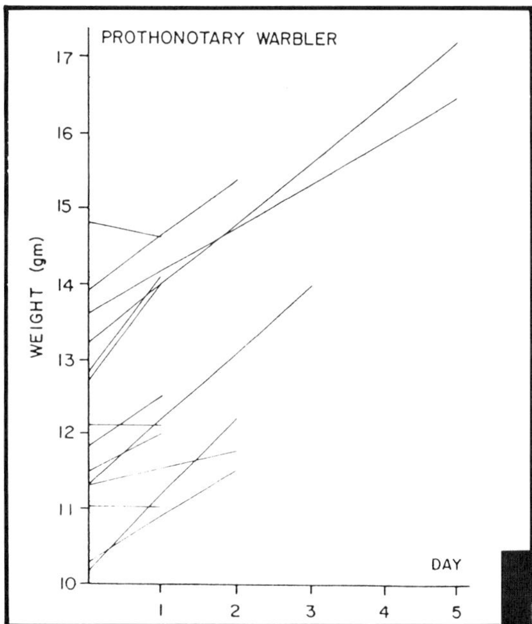

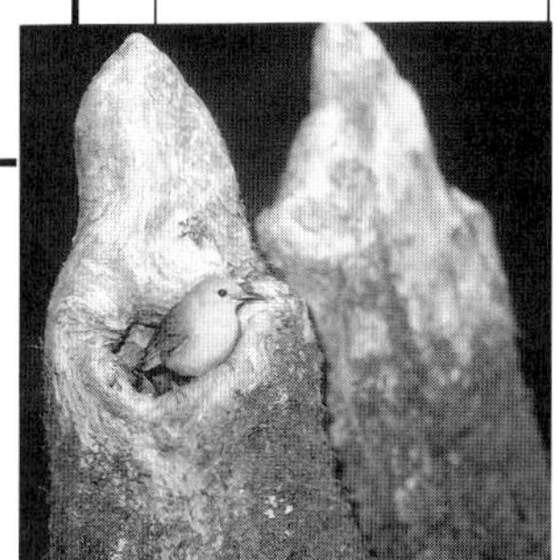

DAVE WARD

□ Figure 2: Migrating songbirds like this Prothonotary Warbler, that remain in an area for several days, increase their body weight 3 to 5% every day. This is equivalent to a 150-pound person gaining 7½ pounds per day.

in spring. During this time they put on more than 50 percent of their body weight in fat before they continue their migration to their breeding grounds in the Canadian arctic.

Most migrant species make stopovers of one to several days. Species like shorebirds that make long flights over water are known to make stopovers of several weeks during which they fatten considerably. During spring migration hundreds of thousands of Red Knots and other shorebirds spend up to two weeks along the Delaware Bayshore eating horseshoe crab eggs and depositing fat. Without this fat they will not reach their arctic breeding sites.

Raptors and other soaring birds also make stopovers. Bill Clark has found that at Cape May some hawks spend from one day to several weeks resting and feeding before continuing autumn migration. Similarly, Katy Duffy has found that some saw-whet and other owls make stopovers of several days in the Cape May area before continuing autumn migration. These birds do not put on massive fat deposits as do songbirds, shorebirds, and some others; so they must feed almost daily during migration. In addition, the hawks use soaring flight which uses much less energy than flapping flight (and consequently they need not put on massive fat deposits.)

What we know about migratory stopovers and weight gain comes from a small number of species at a smaller number of sites. We simply do not know enough about the behavior and ecology of migrants during stopovers. Suffice it to say that birds must

RED KNOT — ALAN BRADY

□ Red Knot increase their body weight 50% by feeding on horseshoe crab eggs deposited on the beaches of Delaware Bay every May. Without this fuel, the birds could not continue on to the arctic to nest.

LONG-EARED OWL — CLAY SUTTON

□ Migrating owls, like this Long-eared Owl, may remain in Cape May for several days before continuing their migration.

SHOREBIRDS — FRANK SCHLEICHER

□ Though our understanding of migratory behavior and ecology is incomplete, one thing is certain: Birds must stop and feed enroute if they are to succeed. Photo: Semipalmated Sandpipers, Short-billed Dowitcher, and Sanderling feeding frantically.

PETE DUNNE

□ At Cape May Point, and elsewhere along the New Jersey coast, high quality habitat has been lost to development.

stopover and that a minimum of two days and maximum of up to three or four weeks is necessary to deposit the fat reserves necessary for successful migration.

Habitat: The Good, The Bad, and The Nonexistent

Habitat is the single most important need of migrating birds. Without suitable habitat in which to make stopovers, birds cannot forage, rest, and avoid predation. In other words, they will die.

Not all habitats are equal to a migrant. Forest-dwelling birds need forest and shorebirds need shoreline or mudflat. Furthermore, not all beaches or forests are equivalent in their ability to sustain migrants making stopovers.

Cape May Point and the New Jersey barrier islands are examples of sites where high quality habitat for migrants has been decimated or reduced to a fraction of its former extent. To birders these migration "traps" are mecca. Their annual pilgrimages to these areas are as regular as that of the migrating birds. The abundance of migrants in these habitats gives the impression that birds are doing well or that they prefer such habitats. Although they may prefer these sites, it is likely that there is just not enough productive habitat. Two examples will illustrate how habitat is not all equal in its ability to support migrant birds.

Six years of aerial shorebird surveys by the Cape May Bird Observatory and New Jersey

PETE DUNNE

□ Ruddy Turnstones and other shorebirds concentrate where they find the food they need to restore the energy stores depleted by migration.

Department of Environmental Protection (Division of Fish, Game, and Wildlife—Nongame and Endangered Species Program) show that birds that stopover along the Delaware Bayshore do not select the beaches they use in a random manner. Instead, Red Knots, Sanderlings, Ruddy Turnstones, and Semipalmated Sandpipers are clusters on a small proportion of the Delaware Bay beaches. In addition, each species clustered in different places. This pattern changes slightly from year to year, presumably in response to the abundance of horseshoe crab eggs that are deposited in the sand during spring. The horseshoe crabs gather for mating and egg laying along the Bayshore on only a small proportion of the available shoreline and consequently deposit their eggs. Thus, not all beaches are used by the crabs or the shorebirds that depend on horsehoe crab eggs.

My studies of migrant songbirds along the Gulf of Mexico in Mississippi and Louisiana during spring migration reveal that these migrants choose the habitats in which they rest and forage. Their choice occurs within a few minutes after their arrival and also shows a nonrandom selection pattern. Most species of Northern American warblers, vireos, and thrushes gathered in pine forests with brushy understory. These habitats occurred on less than 10 percent of the barrier islands on which I worked, yet greater than 70 percent of the migrants preferred these habitats.

In collaboration with Frank Moore I also found that habitats differed in their ability to sustain migrants. A comparison of the duration of stopovers and weight change patterns between study sites on barrier islands off the coast of Mississippi and coastal woodlands in Louisiana showed that the latter habitats are more suitable for songbird migrants. Warblers, vireos, thrushes, and tanagers remained longer and gained more weight in Louisiana than in Mississippi where they did not gain weight and stayed briefly. This is a result of the difference in quality of the habitats for foraging and resting. In Louisiana an abundance of soft-bodied butterfly and moth larvae ("inch-worms") allowed migrants to gain weight at an extraordinary rate. The same species could not gain weight on Mississippi barrier islands because there was simply little food available for them. Consequently songbird migrants did not stopover there.

The absence of habitat for foraging and resting for some species of migrants may be advantageous to other species. For example, bird-eating raptors such as Peregrine Falcons, Merlins, and Sharp-shinned Hawks can prey more easily on birds that cannot find vegeta-

HORSESHOE CRABS

PETE DUNNE

□ Thousands of horseshoe crabs deposit their eggs on the beaches of Delaware Bay, providing an irreplaceable food supply for migrating shorebirds.

tive cover. During spring fieldwork on islands off the coast of Mississippi, I observed raptors waiting for tired songbirds to come in off the Gulf of Mexico. Because there is little forest or brush available and because they are fatigued, these individuals are easy prey. The same situation exists in areas where forest or brush are limited.

HORSESHOE CRAB EGGS

PETE DUNNE

□ Horseshoe crab eggs. The tiny loaves that feed the multitudes of migrating shorebirds.

Suggested Reading

Blem, C.R. 1980. The energetics of migration. In S.A. Gauthreaux, Jr., ed. *Animal Migration, Orientation, and Navigation,* pp. 175-224, New York: Academic Press.

Kerlinger, P. 1989. *Flight Strategies of Migrating Hawks.* Chicago: University of Chicago Press.

Kerlinger, P. and Lein, M.R. 1986. Differences in winter range among age-sex classes of Snowy Owls (*Nyctea scandiaca*) in North America. *Ornis Scandinavica* 17:1-7

Kerlinger, P. and Moore, F.F. 1988. Atmospheric structure and avian migration. *Current Ornithology* 6.

Moore, F.R. and Kerlinger, P. 1987. Stopover and fat deposition by North American wood-warblers *(Parulinae)* following spring migration over the Gulf of Mexico. *Oecologia* 74:47-54.

Poole, A.F. and Agler, B. 1987. Recoveries of Ospreys banded in the United States, 1914-1984. *Journal of Wildlife Management* 51:148-56.

A Sample of Migration Distance for Birds That Breed in New Jersey

SPECIES	WINTER RANGE	FLIGHT DISTANCE (One-way in Kilometers)*
Northern Harrier	New Jersey - Florida	0 — 1,500 +
Northern Bobwhite	same as breeding	0 (trace?)
Green-backed Heron	Maryland - Florida	100+ — >1,000
Sora	North Carolina - Florida (and southward into islands)	600 — 1,500 +
Piping Plover	North Carolina - Florida	600 — 1,500 +
American Woodcock	New Jersey - Florida	0 — >1,000
Black Skimmer	North Carolina - Florida+	600 — 1,500 +
Yellow-billed Cuckoo	Central and South America	>2,500
Screech Owl	New Jersey	0 +
Chimney Swift	Central and South America	>2,500
Ruby-throated Hummingbird	South Florida - Central and South America	1,500 — >2,500
Eastern Kingbird	Central and South America	>2,500
Wood Thrush	Central America	>2,500
Northern Parula	South Florida - Central America	1,500 — >2,500
Northern Oriole	Carolinas - Central America	600 — >2,500
Indigo Bunting	South Florida - Central America	1,500 — >2,500
Field Sparrow	New Jersey - Florida	0 — 1,500

*A range is given because of the variability in migration distance among individuals of a particular species. The winter range varies from year to year so numbers can change.

NEW JERSEY: ONE GREAT REST STOP ON THE GLOBAL HIGHWAY

Richard Kane
NJAS Director of Conservation

PETE DUNNE

□ The Kittatinny Ridge and Cape May—just two of New Jersey's key migratory bird corridors.

The Mainline State

Why is New Jersey so rich in migratory birds and so critical for their global survival? The answer lies partially in our geography and partially in our geological features.

New Jersey is first of all a coastal state. The Atlantic Ocean poses a physical barrier to many migrating land birds. Accordingly, birds tend to concentrate in large numbers along *both* the Atlantic and Delaware Bay coastlines. This is particularly true in the fall. New Jersey's Atlantic coastline is angled northeast to southwest. Land birds traveling east, southeast, or south ultimately intercept the coast and concentrate along its length.

For some species of birds (most notably hawks), coastlines have a cumulative effect. Birds flying south through New England hit the Connecticut or Long Island coast and find their path blocked by water. They take the western bypass and *then* turn south when they reach New Jersey. So not only does New Jersey have to accommodate its own flow of birds, it must support the bird traffic re-routed out of New England!

Our state's latitude also makes it a key link connecting nesting areas in North America with wintering areas in South America. Birds moving from Eastern Canada must pass through New Jersey to reach tropical destinations. For centuries birds have followed traditional paths provided by New Jersey's great linear geologic features—the Kittatinny Ridge, the Hudson River, and the Delaware River. The Cape May Peninsula, extending southward some eighteen miles to sea, is one of the world's greatest migratory bird bottlenecks. Over three hundred species and many millions of migrating birds arrive in Cape May each autumn. Many remain several days.

If we conceive of migration as a transit system, New Jersey is the hub where the major flight lanes along the coast, the river

OPERATION: FLIGHT PATH

FRANK SCHLEICHER
⑧

FRANK SCHLEICHER
⑨

FRANK SCHLEICHER
⑦

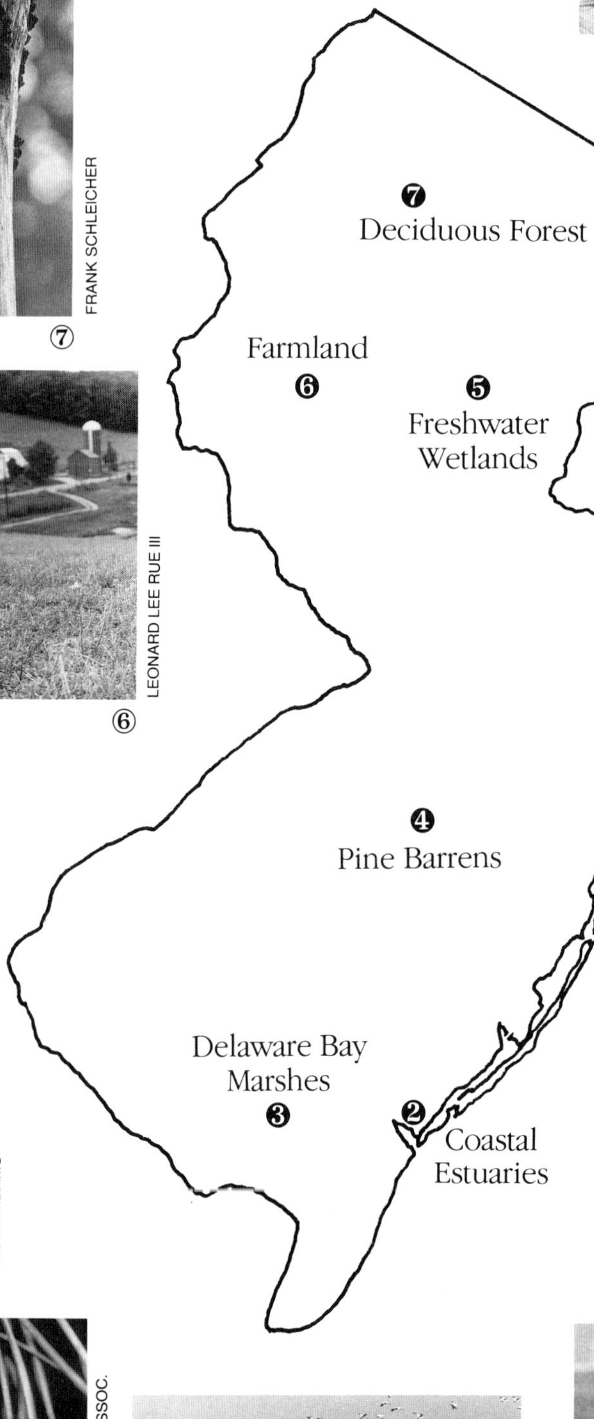

LEONARD LEE RUE III
⑥

NORM ADAMS
⑤

ALAN BRADY
①

ROBERT T. ZAPPALORTI
HERPETOLOGICAL ASSOC.
④

MARION GLASPEY
③

FRANK SCHLEICHER
②

valleys, and the mountains converge and radiate out to destinations north and south. New Jersey is truly the crossroads of migration.

□ New Jersey offers woodland habitat essential to many species of migrating birds (like Hermit Thrush).

HERMIT THRUSH

PETE DUNNE

The Habitat State

One reason why New Jersey supports such an impressive and diverse array of migrants is the variety of physiographic regions and habitats found in our small state. From the ridge and valley provinces of the northwest, to the north central Highlands, to the Piedmont plateau, birds with very different needs find the habitats they require in our state: forests, lakes, marshes, streams, swamps, and grasslands.

In the inner and outer coastal plain, waterfowl and shorebirds find abundant salt marshes and river marshes with rich food supplies, and thousands of woodland birds like warblers, flycatchers, and thrushes seek cover and insects in the hardwood swamps. The beaches and mudflats on Delaware Bay host the largest concentration of horseshoe crabs (and their eggs) in the world, an

PETE DUNNE

□ The second largest spring concentration of shorebirds in North America is found in Delaware Bay each May.

irreplaceable food source for most of the world population of three species of sandpipers.

If we examine the surface area of New Jersey closely and compare the percents and acreage of different habitats, we find some startling facts. About 19 percent of New Jersey's surface is wetlands (the whole United States is only 5 percent); 300,000 acres are salt marsh; 600,000 acres are freshwater wetlands. About 45 percent of New Jersey is forested uplands.

With this kind of habitat diversity, is it any wonder that New Jersey, the "Habitat State," supports such an incredible diversity and number of migrating birds?

A Migratory Calendar

The extent of New Jersey's migration spans the calendar and reaches almost beyond the capacity of human minds to believe it. In early March, you can see a thousand Green-winged Teal on the Maurice River in Cumberland County; in September, at the mouth of the

□ Nineteen percent of New Jersey is wetlands—critical habitat for migrating birds such as the Swamp Sparrow . . .

RICHARD L. DITCH

□ . . . and Green-backed Heron.

SWAMP SPARROW

FRANK SCHLEICHER

PETE DUNNE

□ Forty-five percent of our state is forest, important to Red-shouldered Hawks and many other birds.

RED-SHOULDERED HAWK

FRANK SCHLEICHER

RED KNOT PETE DUNNE

□ In May, Delaware Bay is home for 150,000 Red Knot—virtually the entire New World population.

□ Tree Swallows cloud coastal skies during August and September (and help keep mosquito numbers in check).

very same river, you can witness, from a skiff piloted by a local "bayman," one of the largest concentrations of Sora (rails) in the world.

TREE SWALLOWS ALAN BRADY

In May, on Delaware Bay, virtually the entire North American population of Red Knot (150-200,000) crowd strips of sand like Reeds Beach, Moores Beach and Thompsons Beach. In July at Brigantine National Wildlife Refuge, you can watch 10,000 outbound Short-billed Dowitchers spiral skyward. Destination: Brazil. Every September, 50,000 Broad-winged Hawks sweep across the state. At New Jersey Audubon's hawk lookout in Montclair, you may be treated to the spectacle of 10,000 (maybe even 20,000) Broad-winged Hawks riding the thermals aloft enroute to Central and South America.

Drive along the Garden State Parkway migration corridor to Cape May and you'll see (if not actually count) 100,000 Tree Swallows swirling in clouds or lining wires on the roadsides. When you get to Cape May Point, autumn's Omega Point, you may find a million of them. You'll crane your neck skyward under the shadow of the lighthouse and observe 100 Peregrine Falcons, 200 osprey, 150 Northern Harriers, and 7,000 American Kestrels—*all in an October day!* Stay through the whole season and you can add upwards of 40,000 Sharp-shinned Hawks.

If you watch the ocean off Cape May on an October day, your tally might read 5,000

RED-THROATED LOON PETE DUNNE

□ One of the largest concentrations of Red-throated Loons in the world occurs in Delaware Bay every spring.

FRANK SCHLEICHER

□ Thousands of American Kestrels, a small falcon, migrate down the New Jersey Coast.

OPERATION: FLIGHT PATH

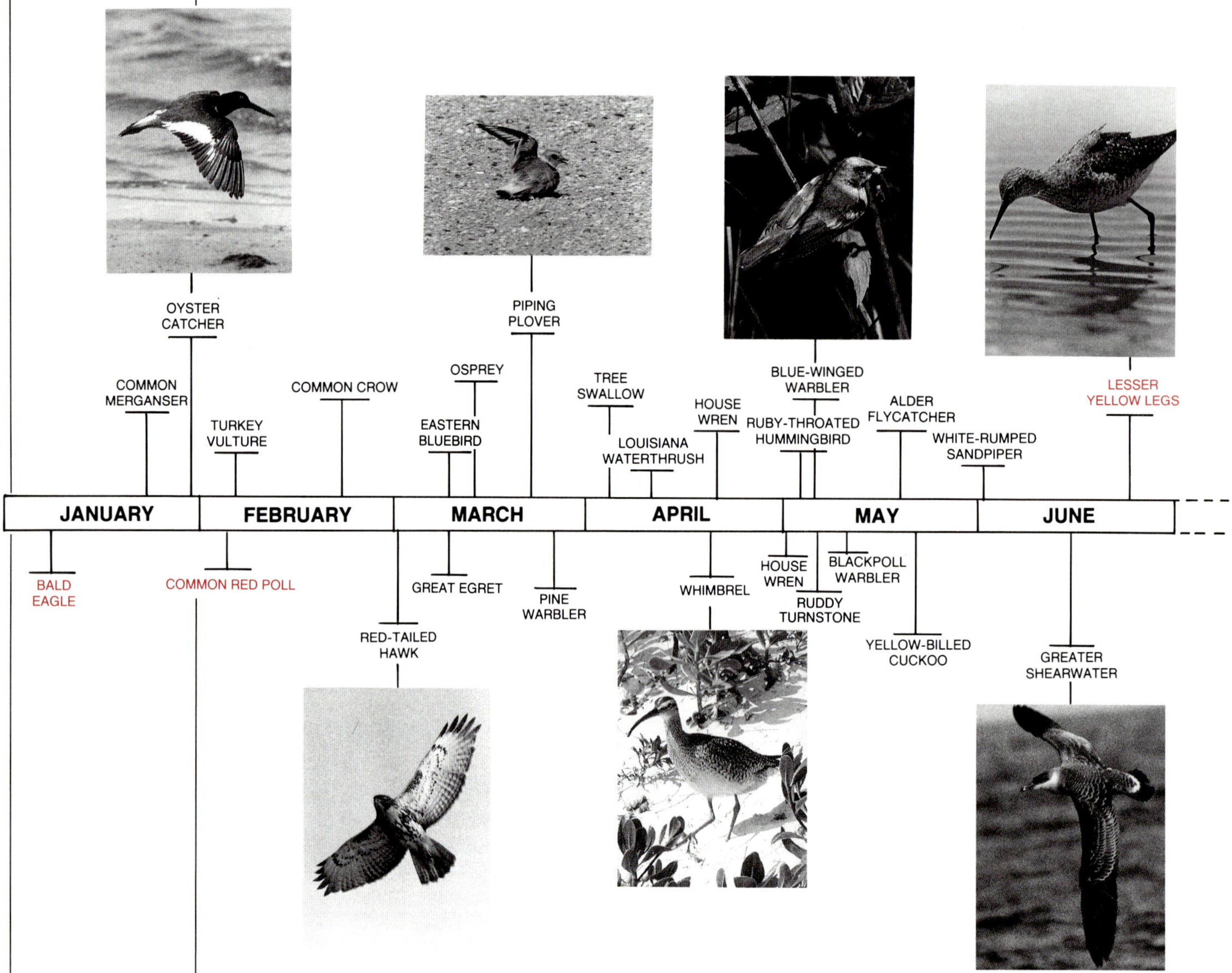

NJ Migration Timetable

(Arrivals Spring and Autumn)

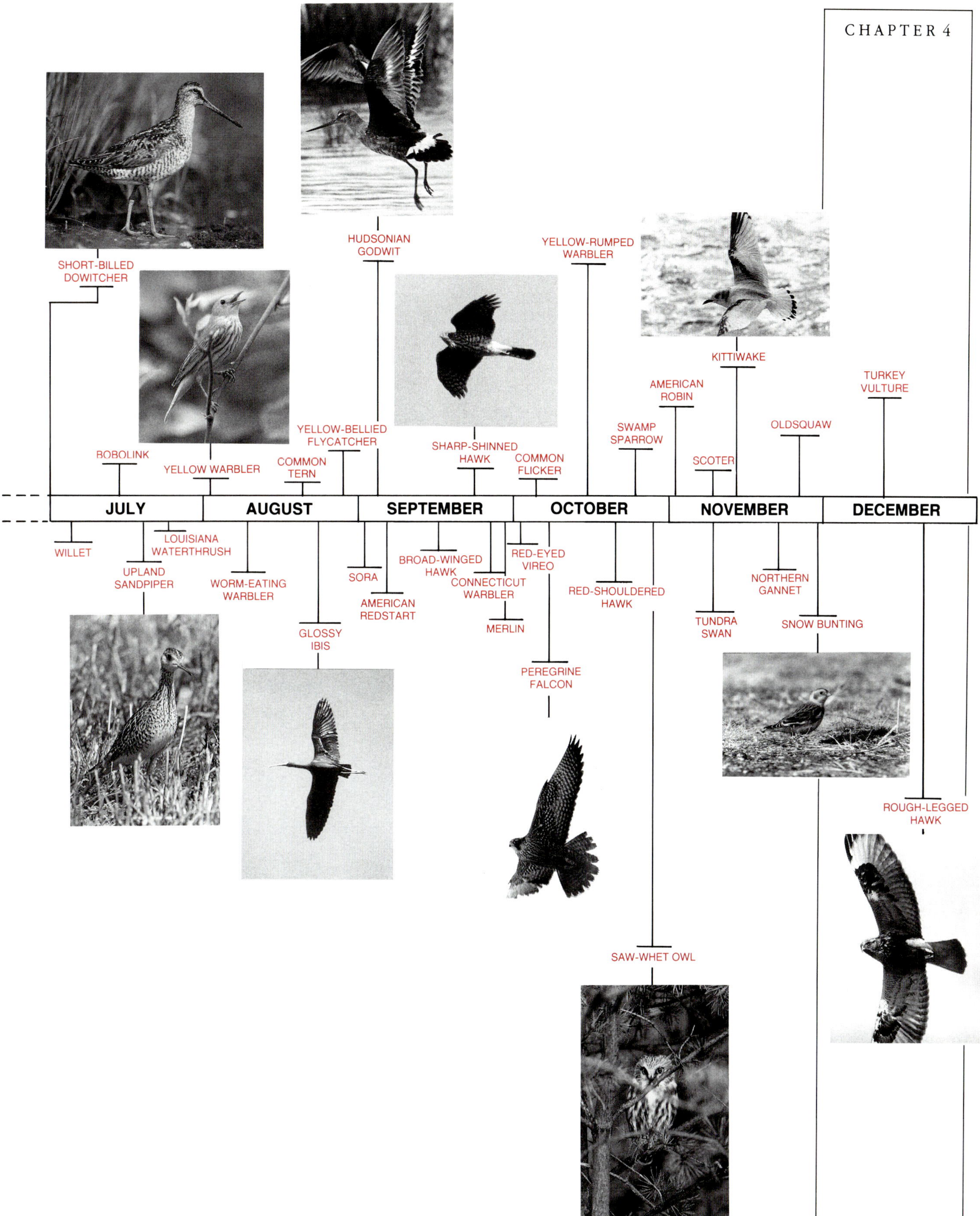
SHORT-BILLED DOWITCHER
HUDSONIAN GODWIT
YELLOW-RUMPED WARBLER
KITTIWAKE
AMERICAN ROBIN
TURKEY VULTURE
SWAMP SPARROW
OLDSQUAW
YELLOW-BELLIED FLYCATCHER
SHARP-SHINNED HAWK
COMMON FLICKER
SCOTER
BOBOLINK
YELLOW WARBLER
COMMON TERN
JULY
AUGUST
SEPTEMBER
OCTOBER
NOVEMBER
DECEMBER
WILLET
LOUISIANA WATERTHRUSH
UPLAND SANDPIPER
WORM-EATING WARBLER
SORA
BROAD-WINGED HAWK
RED-EYED VIREO
CONNECTICUT WARBLER
AMERICAN REDSTART
MERLIN
GLOSSY IBIS
RED-SHOULDERED HAWK
NORTHERN GANNET
TUNDRA SWAN
SNOW BUNTING
PEREGRINE FALCON
ROUGH-LEGGED HAWK
SAW-WHET OWL

□ Different species of birds require different habitat. Destroy the habitat and we jeopardize the existence of birds like the Black-crowned Night-Heron.

Double-crested Cormorants, 10,000 scoters (of three species), and thousands of migrating geese. Return on April 1, and you'll witness the arrival of New Jersey's 4,000 Forster's Terns and the staging of 5,000 or 6,000 Red-throated Loons in Delaware Bay.

Guarding the Habitat State

The sheer number and diversity of birds in our spring and fall migrations depend upon a variety of habitats to ensure safe passage. *Not just any old undeveloped place will do. Different species require different habitat to meet their specific needs.* The cuckoos, warblers, flycatchers, and thrushes stopping here to feed require extensive forest tracts. Shorebirds require mudflats in order to feed during the day, and secluded, elevated bars and beaches to roost at high tide. Ducks, rails, herons, and blackbirds require marshes—some fresh, some salt or brackish.

When these habitats are lost, birds cannot feed and they are forced to move on—to try to find, then crowd into, suitable locations already crowded to their capacity or to move into less suitable areas of marginal benefit. If we protect the habitat we protect the lives of birds. Destroy the habitat, and we jeopardize the existence of birds. It is really that simple.

The New Jersey Audubon Society believes that migratory bird habitat conservation continues to be the weak link in global protection efforts. Unless the needs of migratory birds are met, unless the places they need to rest and feed are protected, North American species will continue to decline.

The purchase of Higbee Beach Wildlife Management Area in 1978 by the New Jersey Division of Fish, Game, and Wildlife; the Nature Conservancy's acquisition of the South Cape May Meadows in 1983; and the designation of the Delaware Bay Sister Shorebird Reserve in 1987 are all models of insight. We want to see more like them. The important locations are already known.

New Jersey Migration Corridors

Since birds have specific requirements, it follows that certain areas are more important than others for different groups of species.

BLACK-CROWNED NIGHT-HERON

RICHARD L. DITCH

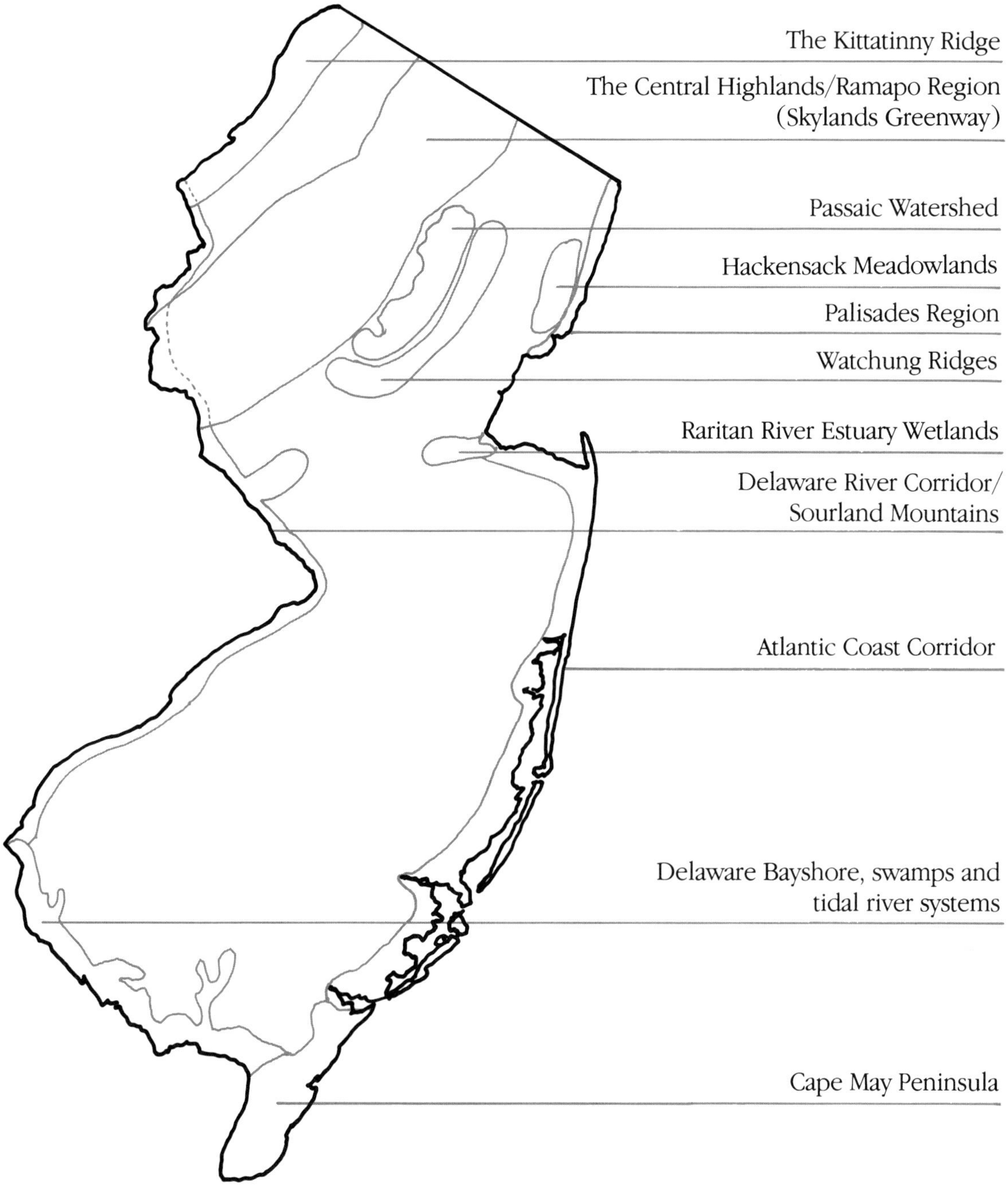

□ The Delaware Bayshore marshes and their adjacent edge rank among the most important migratory bird habitat.

New Jersey Audubon has identified certain key portions of New Jersey that we have designated Migratory Bird Corridors. These areas encompass the Kittatinny Ridge, Pequannock Watershed, Wawayanda Plateau, Hunterdon Plateau, Sourland Mountains, Delaware River Corridor, Palisades Region, Hackensack Meadowlands, all Passaic wetlands, Rockaway Valley, Watchung Ridges, Raritan River Estuary Wetlands, Ramapo Mountain Region, Black River Wetlands, Garden State Parkway Corridor, Atlantic barrier beach and bays and estuaries, Delaware Bayshore and marshes, Cape May Peninsula, Cumberland County swamps and rivers, Atlantic coastal waters to three miles, and Delaware Bay.

PETE DUNNE

Why are These Specific Areas Important?

The Kittatinny Ridge and the Central Highlands (Pequannock Watershed, Wawayanda Plateau) are parts of an ancient ridge system that has been used as a visual guideline for thousands of years by warblers, vireos, flycatchers, thrushes, and raptors. The extensive forests which remain along the steep slopes provide insects and fruit for smaller birds and prey for raptors (hawks). The same is true of the Sourland Mountains and the Hunterdon Plateau.

The Delaware River, as well as the Hudson,

SHORT-EARED OWL

CHRIS STURM

□ A Short-eared Owl hunts over New Jersey's beleaguered Hackensack Meadows—a critical migratory junction made even more important by the loss of habitat around it.

□ Nesting completed, hundreds of American Oystercatcher gather in places like Holgate and Stone Harbor Point in July and August.

Hackensack, and Passaic rivers, concentrate migrants in the same way ridges do. Riverside vegetation offers unbroken woodlands for migrating songbirds. Insect life is plentiful. What's more, associated with these and other rivers (Wallkill, Raritan, Black, Maurice, Cohansey) are rich wetlands that feed thousands of migrating ducks, herons, grebes, rails, snipe and bitterns, as well as assorted migrating songbirds. Examples of these wetlands include: the Hackensack Meadowlands, Trenton Marsh, Troy Meadows, the Great Swamp, and the marshes flanking the Maurice River and its tributaries.

Even isolated woodland pockets along major river corridors can be very important during migration. Some well-known examples are Princeton's Institute Woods and the Glassboro Woods in Gloucester County.

Along the Atlantic Coast Corridor are many critical migratory bird havens just inland from the coastline which some birds follow during migration. Some of these are Allaire and Cheesequake State Parks, Manahawkin swamp, and portions of Bass River and Wharton State Forest, to name a few.

New Jersey's barrier islands and the bays and marshes behind them are accounted among the most famous migratory bird corridors in the world.

The Cape May Peninsula is the single, greatest concentration point in New Jersey—a giant funnel that constricts migration to a narrow gauntlet through which millions of birds must pass. The recent Fish and Wildlife Service plan for a 15,500 acres refuge in Cape May County is significant, timely and commendable.

The Cumberland County hardwood swamps like Bear Swamp and the Maurice-Manumuskin River drainage host some of the largest numbers of spring migrants reported in New Jersey. Some 20,000 waterfowl (Snow Geese, Black Duck, Mallard, Northern Pintail, Green-winged Teal) gather in late winter through early spring in the marshes. Large numbers of raptors—Red-tailed Hawk, Northern Harrier and Bald Eagle—concentrate in the marshes to hunt. Typical numbers of perching birds on a late April morning along a single two-mile stretch of road through Bear Swamp read: 30 Red-eyed Vireo, 15 Yellow-throated Vireo; 25 Black-and-white Warbler; 15 Hooded Warbler; 20 Scarlet Tanager; 35 Chipping Sparrow; 30 Blue-gray Gnatcatcher; 35 Ovenbird; 15 Wood Thrush; 25 Gray Catbirds, and a host of orioles, grosbeaks, and other newly arrived spring migrants.

One of the greatest crises facing New Jersey is how to protect these critical areas that migrating, breeding, and wintering birds depend upon. We have a legacy here to pass on to our children and theirs. We are the last generation who will have this option. If we fail to act, then we have failed them. Because the legacy will be gone.

AMERICAN OYSTERCATCHER

FRANK SCHLEICHER

BALD EAGLES

FRANK SCHLEICHER

□ One pair of Bald Eagles nest in New Jersey; less than 40 winter along our rivers and coasts. Many, many more use New Jersey to rest and feed during migration.

SILENT SPRING; EMPTY AUTUMN

Twenty years ago, this chapter would have been written differently. It would have dealt with *potential* threats to New Jersey's migrating birds. It might have predicted declines in their numbers, might even have offered qualified indications of decline. Quite frankly, this chapter *should* have been written twenty years ago, when these warnings would have been meaningful and timely. Well, it wasn't written. Now the focus has shifted. Now, this chapter deals with loss, not decline, *loss.* Populations of many species that *just twenty years ago were common* have been shattered—reduced in many places to remnants... or memories.

As difficult as this may be to accept, unfortunately, it is quite real. This is the future that everyone dreaded, but nobody could quite bring themselves to believe. This chapter examines the evidence and takes stock of what has been lost. This chapter is the coroner's report.

Shorebirds

At the International Council for Bird Preservation (ICBP) 1986 Conference, Marshall Howe of the U.S. Fish and Wildlife Service and Brian Harrington of Manomet Bird Observatory presented data for twelve species of shorebirds that migrate down the Atlantic Coast of North America. Between 1972-83, three species showed insignificant increases; nine species declined—four significantly: Black-bellied Plover, Whimbrel, dowitcher (mostly Short-billed) and Sanderling. During this period dowitcher declined 40 percent; Whimbrel 60 percent, Sanderling 80 percent. (Source: Dr. Brian Harrington; Dr. Marshall Howe—personal communication.)

SANDERLING

FRANK SCHLEICHER

All three of these species make extensive use of New Jersey's coastal wetlands and beaches in spring and fall. Sixty percent of North America's Sanderling population is estimated to use Delaware Bay each May. Their strategy for survival involves long migrations, involving many thousands of miles with only two or three stops en route. These key staging areas are places where birds rest and feed in preparation for their next long-distance jump. The disruption of even one key link may be disastrous for the species.

□ Delaware Bay is a critical migratory link for 60% of North America's Sanderlings. Its protection is essential if this rapidly declining species is to be maintained.

OPERATION: FLIGHT PATH

□ The number of Sharp-shinned Hawks recorded in Cape May has been declining since 1982. The 20,911 Sharp-shinneds recorded in 1988 is the lowest number since counts were initiated in 1976.

□ At Hawk Mountain, Pennsylvania, concern is growing for the Broad-winged Hawk—a forest raptor whose migration carries it to Central America.

SHARP-SHINNED HAWK
FRANK SCHLEICHER

PEREGRINE FALCON
FRANK SCHLEICHER

BROAD-WINGED HAWK
FRANK SCHLEICHER

□ Cape May has the greatest autumn concentrations of many different species of hawks—including the endangered Peregrine Falcon. Most of these birds are immature and many, perhaps *most,* will not survive.

Hawks

Whereas the populations of most species of hawks have rebounded during the 1970s and 1980s in response to the curtailed use of DDT, two species of forest raptors have turned against the trend.

At Cape May, New Jersey, the number of migrating Sharp-shinned Hawks, which prey upon small birds, has declined alarmingly since 1982—a 40 percent drop in seven years. (Dunne, Peter and Clay Sutton. 1986. Population Trends in Coastal Raptor Migrants over Ten Years of Cape May Point Autumn Counts. *Records of New Jersey Birds,* Vol. 12, No. 3, Autumn [updated by Dunne, August 1989]).

At Hawk Mountain, Pennsylvania, the population of Broad-winged Hawk, a forest species that winters in Central America, has *not* increased. In fact, data shows a decline (as yet, statistically insignificant) that portends trouble if it continues. A species showing significant decline at Hawk Mountain is American Kestrel, a small falcon. (Source: Jim Bednarz, Hawk Mountain Sanctuary, personal communication.)

More birds of prey migrate through New Jersey than any state except Texas. Cape May has the greatest autumn concentrations of Sharp-shinned Hawk, Cooper's Hawk, Red-shouldered Hawk, Northern Harrier, Osprey, Peregrine Falcon, Merlin, and American Kestrel in North America. Ninety percent of these birds are immature and inexperienced which magnifies their need for sufficient habitat.

RED-EYED VIREO

CHRIS STURM

Passerines

Whereas birds that live in New Jersey all year are stable or increasing, declines in the ranks of birds that migrate long distances to spend our winter in the neotropics (the "neotropical migrants") are epidemic.

In the William L. Hutcheson Memorial Forest, New Brunswick, a sixty-acre tract *uncut* since 1700, Dr. Charles F. Leck noted marked declines in breeding neotropical migrants. The Red-eyed Vireo and the Ovenbird, once abundant, *have disappeared as nesting species.* Yellow-throated Vireo, Black-and-white Warbler, and Chestnut-sided Warbler (once common New Jersey breeding birds) have also disappeared. Great Crested Flycatcher, Eastern Wood Pewee, Veery, Northern Oriole, Scarlet Tanager, and Indigo Bunting have declined. (Leck, Charles F. 1982. Declines in Some Forest Birds over Twenty Years. *Records of New Jersey Birds,* Vol. 8 No. 1.)

Species which have increased or remained stable were mostly permanent residents or short-distance migrants. (Leck et al., 1981 cited in Serrao, John. 1985. Decline of Forest Songbirds. *Records of New Jersey Birds,* Vol. 11, No. 1.)

OVENBIRD

FRANK SCHLEICHER

□ In New Brunswick's Hutcheson Memorial Forest, the Red-eyed Vireo and Ovenbird have disappeared as breeding birds.

In Green Brook Sanctuary, a 165-acre mixed-woodland tract atop the Palisades in Tenafly and Alpine, New Jersey, surveys of breeding birds have been conducted since 1949. A comparison of the 1957 and 1983 breeding bird census showed that declining

DAVE SIBLEY CHESTNUT-SIDED WARBLER

JOHN KRICHER LOUISIANA WATERTHRUSH

JOHN KRICHER OVENBIRD

DAVE SIBLEY AMERICAN REDSTART

□ At Green Brook Sanctuary in Bergen County, declining species outnumber birds that are stable or increasing by 2:1. Declining birds include such long-distance warbler migrants as Chestnut-sided Warbler, Ovenbird, Louisiana Waterthrush, and American Redstart.

species outnumbered species showing population increases by almost 2:1. Those species with the most dramatic declines were:

BREEDING BIRD CENSUS

Species	1957	1983
Ruby-throated Hummingbird*	4	0
Eastern Phoebe	10	1
Eastern Wood Peewee*	14	3
Veery*	26	5
Yellow-throated Vireo*	8	0
Red-eyed Vireo*	27	6
Black-and-white Warbler*	24	7
Blue-winged Warbler*	8	0
Black-throated Green Warbler*	7	0
Chestnut-sided Warbler*	11	0
Ovenbird*	27	0
Louisiana Waterthrush*	6	0
Common Yellowthroat*	7	0
Hooded Warbler*	26	0
American Redstart*	25	0
Brown-headed Cowbird	9	0
Field Sparrow	7	0

*denotes a long-distance, neotropical migrant species
Ibid.

In Rock Creek Park, Maryland, an undisturbed woodland tract in the District of Columbia, the nesting migratory songbird population declined 70 percent between 1948 and 1983. Ovenbirds declined by 94 percent; Red-eyed Vireos by 79 percent. (Luoma, Jon. 1988. Nation's Suburbs Blamed for Songbird Decline. *New York Times*, June 21.)

Declines in forest species in Washington, D.C. and Green Brook Sanctuary and many other parts of the suburbanized northeast have been attributed to the fragmentation of woodlots. Birds attempting to breed in small woodland parcels have been shown to be more vulnerable to parasitism by cowbirds and predation from grackles, bluejays, and cats than birds nesting in large, expansive woodland tracts. These species infiltrate woodland tracts via the edge. Large tracts have a higher proportion of core woodlands to edge and less predation and nest parasitism.*

THE MESSAGE IS: CUTTING WOODLANDS TO CREATE EDGE HABITAT OR INFILTRATING WOODLANDS WITH ROADS, POWERLINES, AND DEVELOPMENTS MAKES BREEDING BIRDS VULNERABLE TO PREDATION AND COWBIRD PARASITISM.

**Cowbirds may lay as many as twenty eggs per year in the nests of other birds (neotropical migrants, by comparison, commonly produce only one nest with two to four eggs per season). The larger cowbird chicks outcompete the young of the parenting birds for food.*

In a near virgin hardwood tract located in the Cheat Mountains of West Virginia, George

PETE DUNNE

□ The decline in North America's migrating songbirds has been attributed to the fragmentation of woodlands caused by suburbanization.

□ Declining numbers of songbirds have also been linked to the destruction of tropical rain forest where many winter. Denuded mountainsides show the loss of rain forest in Panama.

DAVE WARD

Hall (1984) found declines over a thirty-five-year period in both the number of species and the breeding density of neotropical migrants. This habitat was unfragmented. Cowbird parasitism "was not a serious factor," yet breeding birds still declined. Between 1947 and 1983:

- Total Number of Species in the study area fell from 25 to 15
- Total Number Neotropical Species fell from 14 to 8
- Total Density* fell from 325 singing males to 273
- Total Density* of Neotropical species fell from 238 to 150

  *singing males/100 acres

(Hall, George A. 1984. Population Decline of Neotropical Migrants in an Appalachian Forest. *American Birds,* Vol. 38, No. 1.)

Songbird declines have also been attributed to loss of wintering habitat in the tropics. It has been estimated that the amount of forest in Central America and the West Indies, where many of our neotropical migrants spend our winter, has been reduced by 50 percent. (Myers, N. 1980. *Conversion of Tropical Moist Forest.* National Academy of Science, Washington D.C.)

David Wilcove, studying large undisturbed woodland tracts in the Great Smoky

DAVE WARD

□ Rain forest destruction begins with road construction that permits access into isolated regions.

Smoky Mountains. (Wilcove, David S. 1988. Changes in the Avifauna of the Great Smoky Mountains: 1947-1983. *Wilson Bulletin,* Vol. 100, No. 2.)

Bruce Ambuel and Stanley Temple, comparing surveys conducted on two large, high quality forest tracts in Wisconsin in 1984 and 1979, found that neotropical migrants had decreased from 81 percent to 43 percent of the total breeding individuals in one tract and from 72 percent to 50 percent in the second. Species that increased were medium-distance migrants like Red-winged Blackbird, Common Grackle, House Wren, and American Robin—all very common suburban species. Species that declined were American Redstart, Cerulean Warbler, Ovenbird, Least Flycatcher, Northern Oriole, Scarlet Tanager, Blue-gray Gnatcatcher, and Yellow-throated Vireo—all long-distance migrants.

JOHN KRICHER

JOHN KRICHER

□ Lumbering . . . and cattle ranching complete the destruction.

Mountains, found no significant changes in the populations of neotropical migrants between 1947 and 1983—supporting the premise that forest fragmentation in the northeast, not destruction of habitat in the tropics, was the principal factor governing population declines.

Wilcove cautioned that a decline in neotropical species may have been masked by "accumulated small biases in (the) sampling methods" or the possibility that birds from surrounding marginal territories were filling vacancies in the higher quality habitat in the

Smaller forest tracts surveyed in 1977, 1978, and 1979 had fewer species and a lower percentage of individuals that were forest-dwelling, long-distance migrants. (Ambuel, Bruce and Stanley A. Temple. 1982. Songbird Populations in Southern Wisconsin Forests: 1954 and 1979. *Journal of Field Ornithology,* Vol. 53, No. 2.)

The findings of Ambuel and Temple support the fragmentation argument but also

LEAST FLYCATCHER
DAVE WARD

support the supposition that other factors are involved in the decline of migrating woodland birds. They urge "research on the ecology of forest-dwelling, long-distance migrants during migration and on their wintering range."

So among shorebirds, numbers are declining and the ranks of three species have fallen precipitously low. Breeding habitat is not thought to be a factor. Efforts to understand the rapid decline focus on the places where the birds winter and concentrate during migration.

Within the ranks of birds of prey, the story has been one of recovery—but several woodland species have fought the trend and show early evidence of decline. One, the Broad-winged Hawk, winters in Central America, where tropical deforestation exceeds 50 percent. Another, the Sharp-shinned Hawk, a bird that preys upon woodland passerines, is also declining.

HOUSE WREN
ARTHUR PANZER

□ Studies in Wisconsin disclose that long-distance migrants like the Least Flycatcher are declining; whereas common suburban species, such as House Wren, increased.

OPERATION: FLIGHT PATH

□ Banding studies conducted by Cape May Bird Observatory found that woodlands attracted greater numbers and the greatest diversity of migrating birds than adjacent fields and marsh.

Among migrating woodland songbirds, decline is widespread. Loss of habitat and associated fragmentation of woodlands is regarded as a prime cause of decline. Loss of habitat in the tropics is another.

As with shorebird decline, the growing challenges that songbirds face during their migration is suspected to be a factor contributing to their decline. The loss of migratory habitat is looked upon as a causal agent. Unfortunately, studies focused upon the habitat needs of migratory songbirds are few but, as the following study shows, these needs are specific.

Between 5 September and 29 October 1985, Mary Gustafson studied habitat use among migrating birds in Cape May. Of the 1,560 birds of 68 species captured and released, the greatest number (619) and the greatest diversity of species (54) were captured in woods. Nets strung in field habitat accrued only 264 individuals of 28 species. Twenty-five species were captured *exclusively* in the woods as compared to four species in the fields. (Gustafson, Mary. 1986. Passerine Habitat Use at Higbee Beach Wildlife Management Area, 1985. *Records of New Jersey Birds,* Vol. 12, No. 4.)

DAVE WARD

Woodland is evidently more important to more passerine species of birds during migration; fields and open areas which are a by-product of suburbanization are used by fewer species and numbers. "Many species of birds are specialized in their habitat requirements and this habitat specificity holds when they are migrating." What happens when birds do not find proper habitat or favorable conditions during migration is unmeasured but the effects are frequently manifest.

On the night of 22-23 September 1983 a tremendous wave of migrating birds moved through Cape May, New Jersey, spurred by a strong cold front. All available habitat was saturated with feeding birds. More impressive and disturbing were the number killed.

In front of the Grand Hotel in Wildwood Crest, two Northern Waterthrush and an Ovenbird were picked up in front of the hotel's glass door—killed by striking the glass. In answer to a call from a resident in Villas, on the other side of the peninsula, thirteen birds were collected from beneath the telephone wires in front of two suburban residences (four Swainson's Thrush, four Ovenbirds, and five Northern Waterthrush).

There is no evidence that the number of birds that were killed by striking windows and roadside wires was uniform along the length and breadth of the Cape May peninsula, but, by the same token, there is no compelling reason to believe otherwise. The number of birds that died on that single heavy flight night in Cape May though impossible to estimate was undeniably staggering. Flights of this magnitude occur two or three times every year.

They used to happen more often than this.

CHAPTER 6

SAVING SEATS IN A CROWDED ROOM

SNOWY OWL

FRANK SCHLEICHER

There is nothing incompatible about birds and a prosperous, thriving New Jersey. Quite the contrary. Eroding the environmental quality of our state will ensure bankruptcy in our

□ A Snowy Owl, an annual migrant and winter resident in New Jersey, finds a bit of peace at Liberty State Park. In the background, the World Trade Center.

SCAUP

FRANK SCHLEICHER

□ Scaup in flight over New York Harbor—an important wintering and migratory stopover point for many migratory birds.

SNOW GEESE

DAVE WARD

□ Snow Geese and tens of thousands of other waterbirds make use of Forsythe National Wildlife Refuge just north of Atlantic City.

NORM ADAMS

□ New Jersey residents engaged in one of New Jersey's most popular outdoor pursuits—enjoying the simple pleasure of being in a natural area.

spirit and our endeavors. As the stewards of this planet, the responsibility for its course lies with us. For the sake of its survival and for the sake of our survival, it is incumbent that we maintain the natural diversity around us.

This task includes maintaining the complex and fragile global pattern of bird migration—important for the birds themselves; important for the land and habitat that will be preserved for them (and for us). Accordingly, the needs of migrating birds *must become part of our planning process.* Our plans must take bird migration into account, and our actions will need to accommodate their needs.

We can, of course, choose the short-sighted path. We can abrogate our stewardship and destroy, by design, default, or expedience, our natural resources for selfish or short-term benefits. That path leads inexorably to a sterile world where ocean rot, nothing blooms, and no bird sings. It leads to extinction. No serious business adopts a strategy of short-term gain followed by bankruptcy. No environmental policy should either.

NORTHERN HARRIER

FRANK SCHLEICHER

□ An adult Northern Harrier hunting the Hackensack Meadows. Perhaps 30 pair nest in our state. Another 750, or so, winter. But several thousand make use of our state twice each year during migration.

Recognizing and ensuring the needs of migrating birds is one important step *away* from that bankrupt path—a necessary step. This chapter offers a Master Plan for protecting migrating birds in New Jersey. It provides important and necessary measures to ensure that this objective is reached. If the needs of migrating birds are to be met, and if the decline is to be halted, these are the things that must be done. It is that simple.

I. Protection of New Jersey's Migration Corridors

Because of natural and topographic features, because of existing patterns of development, and because of the nature of bird migration, there are key migration corridors that birds have come to depend upon. *Conservation strategies and guidelines to assist developers should be drafted to ensure that the integrity of these key corridors is maintained.*

RICHARD L. DITCH

□ If we destroy our environment, we abrogate our obligation as stewards. That course leads to extinction.

A. NEW JERSEY'S CRITICAL MIGRATION CORRIDORS ARE:

1. The Kittatinny Ridge from New York to the Delaware Water Gap
2. The Central Highlands/Ramapo Mountain Region (Skylands Greenway)
3. Delaware River Corridor
4. Palisades Region
5. Hackensack Meadowlands
6. Passaic Watershed
7. Watchung Ridges/Sourland Mountains
8. Raritan River Estuary Wetlands
9. Atlantic Coast Corridor
10. Delaware Bayshore marshes, swamps, and tidal river systems
11. Cape May Peninsula

B. GUIDELINES FOR THE PROTECTION OF CRITICAL MIGRATION CORRIDORS SHOULD INCLUDE THE FOLLOWING PRINCIPLES AND SPECIFIC ACTIONS.

1. As much land as possible should be acquired and placed in public trust either through state, federal, or private conservation agencies to ensure permanent preservation.

2. Development within migration corridors must take into account the needs of migrating birds. Guidelines must be established to serve this end.

SPECIFICALLY:

a. *The integrity of large woodland tracts must be maintained.*

The larger the woodlot, the more useful and productive it is for migratory birds. Parceling woodlands into smaller lots means: (1) Fewer woodland birds will be able to nest; (2) disturbance of nesting birds by people will increase; (3) more nests will fail because of cowbird parasitism, predation, and competition with edge species.

(1) Cluster housing is preferable to parceled lots. Housing units should be localized within large tracts of land and the balance of the land should be maintained as natural land to be used only for passive recreation.

□ Habitat lost to development places a premium and an added burden on those natural areas that remain.

PETE DUNNE

□ Eastern Bluebirds can be encouraged to nest in large fields and pastures by erecting nesting boxes—but maintaining the proper habitat is essential.

EASTERN BLUEBIRD

ARTHUR PANZER

BOBOLINK

O.S. PETTINGILL/VIREO

□ Bobolinks, whose migration will carry them all the way to Argentina, can be encouraged to live in large fields and pastures if mowing is delayed until young are fledged in July.

(2) If cluster housing is not feasible, structures should be arranged around the perimeter of large tracts to minimize destruction of the core area.

b. *Structures should never be situated on ridgetops or within or adjacent to wetlands.*

Ridgetops act as natural leading lines, concentrating migrating birds. Ridgetop structures disrupt migration and kill migrating birds through striking windows, power lines and towers. *Wetlands* are particularly productive in terms of food and cover and vitally important to migrating birds.

c. *Large plate glass windows should be avoided when building along a known flight path.*

Houses should sit so that large windows do not sit at a right angle to the flight (in most places, north and south). In locations like Cape May, New Jersey, where massive concentrations of migrating birds occur regularly, large plate glass windows and elevated telephone lines should be avoided.

d. *During construction, as little natural habitat as possible should be destroyed.*

3. Tax incentives should be established that encourage and reward landholders

□ During migration, *all* habitat becomes critical habitat, no matter how small or insignificant it may appear, and may sustain long-distance migrants like this Ovenbird.

OVENBIRD

DAVE WARD

(private or corporate) who maintain their property for migratory bird use and passive recreation.

4. Homeowners in migration corridors can enhance the utility of their property for migrating birds by:

a. *Not sanitizing woodlots of understory trees, fallen limbs, leaves, plants, and other woodland debris.*

b. *Planting native plants and fruiting trees that provide fruit or seeds.* (See Appendix B.)

5. During migration, *all* habitat becomes critical habitat—including those in heavily urbanized areas. Even small patches of trees, shrubs, or grasses are used by migrating birds and their protection should be encouraged.

II. Lands Already in the Public Domain. These lands should adjust their management policies to accommodate the needs of migrating birds.

A. MIGRATING BIRDS SHOULD BE GIVEN THE SAME LEVEL OF INTEREST AND RESOURCE ALLOCATION ACCORDED NESTING SPECIES.

The traditional management focus upon protecting breeding species is insular and unrealistic. It fails to recognize New Jersey's obligation to protect birds that may spend relatively little time in our state but *time critical to their survival.* It fails to recognize the importance of New Jersey in a global survival strategy.

B. MIGRATING NONGAME SPECIES SHOULD BE GIVEN THE SAME LEVEL OF INTEREST AND PROTECTION AFFORDED GAME SPECIES.

Of the 832 species of migratory birds covered by four bird-protection treaties, 757 or 91 percent are not ordinarily hunted nor are they listed as Threatened or Endangered under the Endangered Species Act. Yet while millions of dollars are spent annually to benefit hunted species, primarily waterfowl, the U.S. Fish and Wildlife Service now spends less than *one million dollars* on nongame bird research. (Senner, Stanley E. 1986. Federal research on migratory nongame birds: is

□ Songbirds deserve the same interest and resource allocations accorded game species.

the United States Fish and Wildlife Service doing its job? *American Birds,* Vol. 40, No. 3, Fall.)

C. POLICIES AND MANAGEMENT PRACTICES IN NEW JERSEY'S PARKS, FORESTS, WILDERNESS AREAS, NATURAL AREAS, REFUGES AND RECREATION AREAS (AS WELL AS STATE RUN HOSPITALS, PRISONS, AND FEDERAL MILITARY FACILITIES, AND PROPERTY MAINTAINED BY THE DEPARTMENT OF TRANSPORTATION [HIGHWAY INTERCHANGES, CENTER ROADWAY DIVIDES, WIDE SHOULDERS, REST AREAS]) SHOULD TAKE INTO ACCOUNT THE NEEDS OF MIGRATING BIRDS. THESE NEEDS ARE: *FOOD, SHELTER, SPACE.* HABITAT MANAGEMENT PRACTICES ON STATE AND FEDERAL PROPERTY SHOULD BENEFIT MIGRATING BIRDS OR AT THE VERY LEAST NOT JEOPARDIZE THEIR EXISTENCE. SPECIFIC ACTIONS INCLUDE:

□ Water levels in impoundment pools located on federal and state refuges should be drawn down during migration to provide habitat for many thousands of migrating shorebirds.

DOWITCHER AND SANDERLING

FRANK SCHLEICHER

□ Management policies should emphasize diversity. If all natural lands were managed for the American Woodcock (at right), a bird of fields and young woodlands, the Pileated Woodpecker (opposite page), a bird of mature forests, would disappear.

1. State, county and local forest, parks, and wildlife management areas should be inventoried to establish the number and diversity of *breeding, wintering,* and *migrating* birds.
2. Stream corridors should be assiduously protected and maintained.
3. Woodland understory should be uncut and left natural to provide vital habitat.
4. Water levels in impoundment pools within federal and state natural areas should be drawn down in May, July, August, and September to accommodate use by migrating shorebirds, rails, and other species as is now being done in other states.
5. Land adjacent to highways and roadsides should be allowed to revert to natural vegetation or planted with a native cover that provides food and shelter. A prime example of this is Route 287 between Bernardsville and Morristown, New Jersey.
6. A broad, statewide policy containing management guidelines for migrating bird use should be drafted and adopted by all New Jersey departments and county park systems.

D. WILDLIFE AREAS SHOULD BE MANAGED FOR WHATEVER SPECIES ARE THERE; NOT IN MIMICRY OF ANOTHER LOCATION OR SPECIES EMPHASIS.

There is a tendency to impose blanket management policies in every location. Whereas these benefit some target species, they are often detrimental to many others, and usually, they are implemented without proper regard for indigenous plants and animals. As Rosalie Edge once observed: "The time to save a species is while it is still common." *Management practices should look to maintain New Jersey's statewide natural diversity which includes migrating birds.* If, for example, all of New Jersey

AMERICAN WOODCOCK

FRANK SCHLEICHER

PILEATED WOODPECKER

FRANK SCHLEICHER

were managed for the edge-loving woodcock, a bird of mature fauna, the pileated woodpecker (and many other species) would disappear.

E. WILDLIFE MANAGEMENT SHOULD AIM TOWARD SPECIES COMMUNITIES (E.G., FOREST BIRDS, GRASSLAND BIRDS, COLONIAL BREEDING BIRDS, SHOREBIRDS) AND NOT TOWARD SINGLE SPECIES MANAGEMENT AS IS PREVALENT TODAY.

Management programs are often directed to benefit individual species. Usually, but not always, these are game species. Usually, the target species is declining and management is perceived as a way of bolstering its numbers. But these management policies are often detrimental to a much larger number of nontarget species—many of which may also be in decline.

F. THE EFFECTS OF MANY MANAGEMENT PRACTICES UPON MIGRATING BIRDS ARE NOT KNOWN OR ARE POORLY UNDERSTOOD AND NEED TO BE INVESTIGATED.

PETE DUNNE

□ The common and widespread practice of applying chemical poisons to control insect pests has a direct and negative impact on migrating *and* nesting birds.

□ Aerial spraying to control gypsy moths in Stokes State Forest was shown to reduce the number of nesting birds like the Great Crested Flycatcher (right) and the Black-and-White Warbler (opposite page) by over 50%.

1. The effect of open marsh management and other mosquito control practices upon migratory shorebird utilization *and* native salt marsh breeding birds (e.g., Northern Harrier and Seaside Sparrow).
2. The value of certain widely used cover crops for migrating birds.
3. Woodcutting practices including the removal of standing dead timber.
4. The policy of cutting, poisoning or burning *Phragmites communis* or foxtail. There is mounting evidence that this plant is very heavily utilized by migrating songbirds and rails.

III. Insect Control Activities

The practice of widespread aerial spraying of chemicals to control insect pests has a direct, negative impact on migrating and nesting birds. Aerial spraying during May (to control gypsy moths) and in July-October (to control mosquitoes) is not specific to the target insects. It eliminates the food that birds depend upon in migration. Next to habitat

GREAT CRESTED FLYCATCHER

DAVE WARD

destruction, aerial spraying remains the single greatest threat to migrating birds. In the aftermath of DDT, we switched chemicals, and changed nothing. To protect and provide for birds that breed and migrate through New Jersey it is incumbent that:

A. THE COSMETIC MASS AERIAL SPRAYING FOR GYPSY MOTH CONTROL DURING MAY AND JUNE SHOULD BE STOPPED.

1. A 1971 study in Stokes State Forest disclosed that one aerial application of one of the most widely used insecticides reduced bird abundance by 55 percent. Bird populations continued to be 45 percent lower than unsprayed plots the following year. (Moulding, Jonathan D. 1976. Effects of a Low-Persistence Insecticide on Forest Bird Populations. *Auk,* Vol. 93, No. 4.)
2. Biological alternative insecticides, touted by proponents as a safe substitute for chemicals because they are specific to "leaf-eating caterpillars," still kill many species of nontarget caterpillars that many species of birds depend upon for food.
3. The first outbreaks of gypsy moths occurred in Massachusetts shortly after 1868, long before the advent of aerial spraying. The forests are still standing.
4. Prudent, localized spraying for gypsy moth control by property owners to protect

BLACK-AND-WHITE WARBLER
PETE DUNNE

BLUE-WINGED WARBLER
FRANK SCHLEICHER

HAIRY WOODPECKER
FRANK SCHLEICHER

shade trees does not have the same devastating effects as blanket control efforts.
5. Though unwanted, and a perfect example of the sad consequences of exotic introductions, the gypsy moth has become part of our state's natural web. In time, tree species that are most vulnerable to gypsy moth deforestation will be replaced by more resistant species. Trees that weaken and die are still important components to the woodland, providing food and shelter for a number of cavity nesting species.

□ Caterpillars are a prime food for many species of migratory birds—including the Blue-winged Warbler.

EASTERN BLUEBIRD
FRANK SCHLEICHER

□ Dead trees provide nesting cavities for many species including Hairy Woodpecker and Eastern Bluebird.

B. THE WIDESPREAD AERIAL SPRAYING OF ADULTACIDE AND LARVACIDE TO CONTROL MOSQUITOES IN FRESH WATER AND TIDAL MARSHES BE DISCONTINUED BETWEEN MAY-OCTOBER.

1. The most commonly used mosquito control agents are not specific to adult mosquitoes or mosquito larvae. Aquatic animals that provide food for migrating shorebirds are also killed.
2. In areas where mosquitoes pose a clear health (not nuisance) threat, control efforts should be preventive instead of reactive. Concerted control efforts in winter and early spring (March/April) when mosquito numbers are low, localized, and immobile would not conflict with most migratory bird use.

The problem with mosquito control is not a matter of methodology but one of public attitude. Over most of New Jersey and for most of their breeding season, the insects pose no health threat to humans. They are an annoyance and an inconvenience and for this reason millions of tax dollars are expended and widespread environmental disruption is sanctioned—freshwater wetlands are drained, insecticides are applied. These activities have direct and detrimental effects upon birds—both by altering the habitat many specialized species of birds require or by eliminating the food native and migrating birds depend upon.

Until New Jersey citizens understand and affirm that widespread environmental degradation is too high a price to pay for human convenience, mosquito control activities will continue to threaten the survival of birds dependent upon New Jersey's woodlands and marshes.

NORM ADAMS

□ The phenomenon of migration is understudied and, as a means of bringing New Jersey residents to a better understanding of their environment, it is without peer. Pictured is one of the regular hawk-banding demonstrations sponsored by New Jersey Audubon's Cape May Bird Observatory.

IV. Migration in Education

The phenomena of migration are understudied at every level from grade school through college level wildlife management curricula. Educators can only pass on information that has been brought to their awareness. Migration study has a long history of neglect. We urge:

1. That environmental education be mandatory in teacher colleges and that birds and migration be included in the curriculum.
2. That studies in nongame and migrating birds be increased in environmental, biology, and game management curricula for those seeking degrees and employment in environmental fields.
3. That questions concerning bird migration and nongame birds be included in the civil service tests for positions whose duties will include environmental education or habitat and wildlife management.
4. That migration study be accorded a major place in grammar school science curricula as a means of instilling a sense of such basic concepts as seasonality, habitat, and survival.
5. That bird migration be taught as a unit in basic high school biology and as a means of introducing students into primary research.

APPENDIX A: THE BIRDS OF NEW JERSEY

ABBREVIATIONS

C ***Common*** — frequently encountered or numerous in the appropriate habitat and season.

U ***Uncommon*** — regular but elusive; not always encountered in the appropriate habitat and season.

R ***Rare*** — infrequently to rarely encountered, even in appropriate habitat and season; some are not seen every year or every season.

F ***Former*** — has bred but no longer does so.

HABITAT

FW ***Fresh Water*** wetlands

SW ***Salt Water*** wetlands

SPECIES	BREEDING	MIGRANT	WINTER	PREFERRED HABITAT(S)
Red-throated Loon		C	C	Coastal waters; open ocean
Pacific Loon		R	R	Coastal waters; lakes
Common Loon		C	U	Coastal waters; lakes
Pied-billed Grebe	R	C	U	FW wetlands
Horned Grebe		C	C	Coastal waters; lakes
Red-necked Grebe		R	R	Coastal waters; lakes
Eared Grebe		R	R	Coastal waters
Northern Fulmar		U	U	Open ocean
Cory's Shearwater		C		Open ocean
Greater Shearwater		C		Open ocean
Sooty Shearwater		C		Open ocean
Manx Shearwater		U		Open ocean
Audubon's Shearwater		R		Open ocean
Wilson's Storm-Petrel		C		Opean ocean; coastal waters
Leach's Storm-Petrel		R		Open ocean
Northern Gannet		C	C	Open ocean; coastal waters
American White Pelican		R		SW wetland; coastal waters
Brown Pelican		U		Coastal waters
Great Cormorant		U	U	Coastal waters
Double-crested Cormorant		C	C	Coastal waters
American Bittern	R	U	R	FW, SW wetlands
Least Bittern	U	U	R	FW, SW wetlands
Great Blue Heron	U	C	U	FW, SW wetlands
Great Egret	C	C	R	SW wetlands
Snowy Egret	C	C	R	SW wetlands
Little Blue Heron	U	U	R	SW wetlands

SPECIES	BREEDING	MIGRANT	WINTER	PREFERRED HABITAT(S)
Tricolored Heron	U	U	R	SW wetlands
Cattle Egret	C	C		SW wetlands; fields
Green-backed Heron	C	C	R	FW, SW wetlands
Black-crowned Night-Heron	C	C	U	SW wetlands
Yellow-crowned Night Heron	U	U		SW wetlands
White Ibis		R		SW, FW wetlands
Glossy Ibis	C	C	R	SW wetlands
Fulvous Whistling-Duck		R		SW, FW wetlands
Tundra Swan		C	U	SW, FW wetlands; lakes
Mute Swan	C	C	C	SW, FW wetlands; lakes
Greater White-fronted Goose		R	R	SW, FW wetlands; lakes
Snow Goose		C	U	SW wetlands
Ross' Goose		R	R	SW wetlands
Brant		C	C	Coastal waters; SW wetlands
Canada Goose	C	C	C	FW, SW wetlands; lakes
Wood Duck	C	C	R	FW wetlands
Green-winged Teal	R	C	U	FW, SW wetlands
American Black Duck	C	C	C	SW, FW wetlands
Mallard	C	C	C	FW, SW wetlands; lakes
Northern Pintail	R	C	C	FW, SW wetlands; lakes
Blue-winged Teal	C	C	R	FW, SW wetlands
Northern Shoveler	R	C	C	SW, FW wetlands
Gadwall	C	C	C	Lakes; coastal waters; FW, SW wetlands
Eurasian Wigeon		R	R	Lakes; coastal waters; SW, FW wetlands
American Wigeon		C	C	Lakes; coastal waters; SW, FW wetlands
Canvasback		C	C	Coastal waters; lakes
Redhead		U	R	Coastal waters; lakes
Ring-necked Duck		C	U	FW wetlands; lakes
Greater Scaup		C	C	Coastal waters; lakes
Lesser Scaup		C	C	Coastal waters; lakes
Common Eider		R	R	Coastal waters
King Eider		R	R	Coastal waters
Harlequin Duck		R	R	Coastal waters
Oldsquaw		C	C	Coastal waters
Black Scoter		C	C	Coastal waters; lakes
Surf Scoter		C	C	Coastal waters

SPECIES	BREEDING	MIGRANT	WINTER	PREFERRED HABITAT(S)
White-winged Scoter		C	C	Coastal waters
Common Goldeneye		C	C	Coastal waters; rivers
Barrow's Goldeneye		R	R	Coastal waters
Bufflehead		C	C	Coastal waters; lakes
Hooded Merganser	R	C	U	FW wetlands; lakes
Common Merganser	R	C	U	Lakes; rivers
Red-breasted Merganser	R	C	C	Coastal waters
Ruddy Duck	R	C	C	Coastal waters; lakes
Black Vulture	R	U	U	Mixed uplands
Turkey Vulture	C	C	C	Mixed uplands
Osprey	U	C		SW wetlands; coastal waters; lakes; rivers
American Swallow-tailed Kite		R		Mixed uplands; SW wetlands
Mississippi Kite		R		Mixed uplands
Bald Eagle	R	U	U	SW, FW wetlands; fresh and coastal waters
Northern Harrier	R	C	C	SW, FW wetlands; fields
Sharp-shinned Hawk	R	C	U	Forest
Cooper's Hawk	R	C	U	Forest
Northern Goshawk	R	U	R	Forest
Red-shouldered Hawk	U	C	R	Forest; FW wetlands
Broad-winged Hawk	C	C		Forest
Swainson's Hawk		R		
Red-tailed Hawk	C	C	C	Mixed uplands
Rough-legged Hawk		U	U	SW wetlands; fields
Golden Eagle		U	R	SW wetlands; ridge migrant
American Kestrel	U	C	C	Mixed uplands; fields
Merlin		C	R	SW wetlands; coastal
Peregrine Falcon	R	U	R	SW wetlands
Gyrfalcon		R	R	SW wetlands; fields
Ring-necked Pheasant	C		C	Fields
Ruffed Grouse	C		C	Forest
Wild Turkey	C		C	Forest
Northern Bobwhite	C		C	Fields; thickets
Yellow Rail		R	R	SW, FW wetlands
Black Rail	R	R	?	SW wetlands
Clapper Rail	C	C	R	SW wetlands
King Rail	U	U	R	FW wetlands
Virginia Rail	C	C	U	FW wetlands

SPECIES	BREEDING	MIGRANT	WINTER	PREFERRED HABITAT(S)
Sora	C	C	R	FW wetlands
Common Moorhen	C	C	R	FW wetlands
American Coot	R	C	C	FW wetlands; lakes
Sandhill Crane		R	R	FW wetlands; lakes
Black-bellied Plover		C	C	SW wetlands
Lesser Golden Plover		U		Fields; SW wetlands
Wilson's Plover	F	R		SW wetlands; beaches
Semipalmated Plover		C	R	SW wetlands
Piping Plover	U	U	R	Beaches
Killdeer	C	C	U	Fields; SW wetlands
American Oystercatcher	C	C	R	SW wetlands
Black-necked Stilt		R		SW wetlands
American Avocet		R		SW wetlands
Greater Yellowlegs		C	R	SW wetlands
Lesser Yellowlegs		C	R	SW wetlands
Solitary Sandpiper		C		FW wetlands
Willet	C	C		SW wetlands
Spotted Sandpiper	U	C		FW, SW wetlands
Upland Sandpiper	R	U		Fields
Whimbrel		C		SW wetlands
Hudsonian Godwit		U		SW wetlands
Bar-tailed Godwit		R		SW wetlands
Marbled Godwit		U	R	SW wetlands
Ruddy Turnstone		C	U	Beaches; SW wetlands
Red Knot		C	R	Beaches; SW wetlands
Sanderling		C	U	Beaches
Semipalmated Sandpiper		C		SW wetlands
Western Sandpiper		C	R	SW wetlands
Least Sandpiper		C		SW, FW wetlands
White-rumped Sandpiper		U		SW wetlands
Baird's Sandpiper		R		Fields; SW wetlands
Pectoral Sandpiper		R		SW, FW wetlands
Purple Sandpiper		U	U	Beaches; rock jetties
Dunlin		C	C	SW wetlands
Curlew Sandpiper		R		SW wetlands
Stilt Sandpiper		C		SW wetlands
Buff-breasted Sandpiper		R		Fields
Ruff		R		SW, FW wetlands
Short-billed Dowitcher		C		SW wetlands

SPECIES	BREEDING	MIGRANT	WINTER	PREFERRED HABITAT(S)
Long-Billed Dowitcher		U	R	SW, FW wetlands
Common Snipe	R	C	R	FW wetlands
American Woodcock	C	C	R	FW wetlands
Wilson's Phalarope		U		SW, FW wetlands
Red-necked Phalarope		U		Open ocean; SW wetlands
Red Phalarope		U	U	Open ocean; SW wetlands
Pomarine Jaeger		C	R	Open ocean
Parasitic Jaeger		C	R	Open ocean
Long-tailed Jaeger		R		Open ocean
Great Skua		R	R	Open ocean
South Polar Skua		R		Open ocean
Laughing Gull	C	C	R	SW wetlands; beaches; coastal waters
Little Gull		R	R	Beaches; coastal waters
Common Black-headed Gull		R	R	Beaches; coastal waters
Bonaparte's Gull		C	C	Beaches; coastal waters
Ring-billed Gull	U	C	C	Beaches; coastal waters; lakes
Herring Gull	C	C	C	Beaches; coastal waters; SW wetlands
Iceland Gull		R	R	Beaches; coastal waters
Lesser Black-backed Gull		R	R	Beaches; coastal waters
Glaucous Gull		R	R	Beaches; coastal waters
Great Black-backed Gull	C	C	C	Beaches; coastal waters
Black-legged Kittiwake		U	U	Open ocean; coastal waters
Gull-billed Tern	U	U		SW, FW wetlands
Caspian Tern	R	U		SW wetlands; coastal waters
Royal Tern	R	C		SW wetlands; coastal waters
Sandwich Tern		R		SW wetlands; coastal waters
Roseate Tern	F	R		SW wetlands; coastal waters
Common Tern	C	C		SW wetlands; coastal waters
Arctic Tern		R		Open ocean; coastal waters
Forster's Tern	C	C	R	SW wetlands; coastal waters
Least Tern	U	C		Beaches; coastal waters
Black Tern		R		SW wetlands
Black Skimmer	U	U		SW wetlands
Dovekie		U	U	Open ocean
Common Murre		R	R	Open ocean; coastal waters
Thick-billed Murre		R	R	Open ocean; coastal waters
Razorbill		U	U	Open ocean; coastal waters

SPECIES	BREEDING	MIGRANT	WINTER	PREFERRED HABITAT(S)
Atlantic Puffin		R	R	Open ocean
Rock Dove	C		C	Towns; farmlands
Mourning Dove	C	C	C	Mixed uplands
Black-billed Cuckoo	U	C		Forest; forest edge
Yellow-billed Cuckoo	C	C		Forest; forest edge
Common Barn-Owl	U	U	U	Fields; farms
Eastern Screech-Owl	C		C	Mixed uplands
Great Horned Owl	C		C	Mixed uplands
Snowy Owl		R	R	Beaches; fields
Barred Owl	U		U	FW wetlands; forest
Long-eared Owl	R	U	U	Mixed uplands; conifers
Short-eared Owl	R	U	U	Fields; SW, FW wetlands
Northern Saw-whet Owl	R	U	U	Mixed uplands; conifers
Common Nighthawk	U	C		Mixed uplands
Chuck-will's-widow	U	U		Forest
Whip-poor-will	C	C		Forest
Chimney Swift	C	C		Towns
Ruby-throated Hummingbird	U	C		Mixed uplands
Belted Kingfisher	C	C	U	FW wetlands
Red-headed Woodpecker	R	U	R	Forest
Red-bellied Woodpecker	C		C	Forest
Yellow-bellied Sapsucker		U	R	Forest; orchards
Downy Woodpecker	C		C	Forest; forest edge
Hairy Woodpecker	C		C	Forest
Black-backed Woodpecker		R	R	Forest
Northern Flicker	C	C	U	Mixed uplands
Pileated Woodpecker	U		U	Forest
Olive-sided Flycatcher		U		Forest; forest edge; FW wetlands
Eastern Wood-Pewee	C	C		Forest
Yellow-bellied Flycatcher		U		Forest
Acadian Flycatcher	C	C		Forest
Alder Flycatcher	U	U		FW wetlands
Willow Flycatcher	C	C		FW wetlands; old fields
Least Flycatcher	U	C		Forest
Eastern Phoebe	C	C	R	Forest; forest edge; near water
Great Crested Flycatcher	C	C		Forest
Western Kingbird		R		Mixed uplands
Eastern Kingbird	C	C		Mixed uplands

SPECIES	BREEDING	MIGRANT	WINTER	PREFERRED HABITAT(S)
Horned Lark	U	C	C	Fields
Purple Martin	C	C		FW, SW wetlands; mixed uplands
Tree Swallow	C	C	R	FW, SW wetlands; fields
Northern Rough-winged Swallow	C	C		FW, SW wetlands; fields
Bank Swallow	C	C		FW wetlands; fields
Cliff Swallow	R	U		FW wetlands; fields
Barn Swallow	C	C		Fields
Blue Jay	C	C	C	Forest
American Crow	C	C	C	Mixed uplands
Fish Crow	C	C	C	SW, FW wetlands
Common Raven	F	R	R	Forest
Black-capped Chickadee	C		C	Forest; forest edge
Carolina Chickadee	C		C	Forest; forest edge
Boreal Chickadee		R	R	Forest, conifers
Tufted Titmouse	C		C	Forest; mixed uplands
Red-breasted Nuthatch	R	U	U	Forest; conifers
White-breasted Nuthatch	C		C	Forest
Brown Creeper	U	C	U	Forest
Carolina Wren	C		C	Thickets; forest
House Wren	C	C		Mixed uplands
Winter Wren	R	U	U	Forest; thickets
Sedge Wren	R	R	R	FW, SW wetlands
Marsh Wren	C	C	R	SW, FW wetlands
Golden-crowned Kinglet	U	C	U	Forest; conifers
Ruby-crowned Kinglet		C	U	Forest; thickets
Blue-gray Gnatcatcher	C	C		Forest
Northern Wheatear		R		Fields
Eastern Bluebird	C	C	U	Mixed uplands
Veery	C	C		Forest
Gray-cheeked Thrush		U		Forest
Swainson's Thrush		C		Forest
Hermit Thrush	U	C	U	Forest; thickets
Wood Thrush	C	C		Forest
American Robin	C	C	C	Mixed uplands
Varied Thrush		R	R	Forest; feeders
Gray Catbird	C	C	R	Forest edge; old fields
Northern Mockingbird	C		C	Mixed uplands; suburbs
Brown Thrasher	C	C	R	Old fields; forest edge; hedgerows

SPECIES	BREEDING	MIGRANT	WINTER	PREFERRED HABITAT(S)
Water Pipit		C	U	Fields; dunes
Cedar Waxwing	U	C	U	Forest; forest edge
Northern Shrike		R	R	Mixed uplands
Loggerhead Shrike	R	R	R	Mixed uplands
European Starling	C		C	Universal
White-eyed Vireo	C	C		Old fields; forest edge
Yellow-throated Vireo	C	C		Forest
Solitary Vireo	U	C		Forest
Warbling Vireo	C	C		Forest; near water
Philadelphia Vireo		U		Forest
Red-eyed Vireo	C	C		Forest
Blue-winged Warbler	C	C		Old fields; forest edge
Golden-winged Warbler	U	U		Old fields; forest edge
Tennessee Warbler		C		Forest
Orange-crowned Warbler		R	R	Old fields; thickets; forest edge
Nashville Warbler	R	U		Forest; thickets
Northern Parula	R	C		Forest
Yellow Warbler	C	C		Old fields; forest edge; FW wetlands
Chestnut-sided Warbler	C	C		Old fields; forest edge
Magnolia Warbler	R	C		Forest
Cape May Warbler		U		Forest; conifers
Black-throated Blue Warbler	R	C		Forest
Yellow-rumped Warbler	R	C	C	Old fields; forest edge; dune thickets
Black-throated Green Warbler	U	C		Forest
Blackburnian Warbler	U	C		Forest; conifers
Yellow-throated Warbler	U	U		Forest; near water
Pine Warbler	C	C	R	Forest; conifers
Prairie Warbler	C	U		Old fields
Palm Warbler		C	R	Old fields; shrubby dunes
Bay-breasted Warbler		U		Forest
Blackpoll Warbler		C		Forest
Cerulean Warbler	U	U		Forest
Black-and-white Warbler	U	C		Forest
American Redstart	C	C		Forest; forest edge
Prothonotary Warbler	U	U		FW wetlands
Worm-eating Warbler	U	U		Forest
Ovenbird	C	C		Forest

SPECIES	BREEDING	MIGRANT	WINTER	PREFERRED HABITAT(S)
Northern Waterthrush	U	C		Forest; near water
Louisiana Waterthrush	C	C		Forest; near water
Kentucky Warbler	U	U		Forest
Connecticut Warbler		R		Forest; forest edge; thickets
Mourning Warbler		R		Forest; forest edge; thickets
Common Yellowthroat	C	C	R	Thickets; old fields; FW, SW wetlands
Hooded Warbler	U	U		Forest
Wilson's Warbler		U		Forest edge; thickets
Canada Warbler	U	C		Forest; forest edge
Yellow-breasted Chat	U	U		Old fields; forest edge
Summer Tanager	U	U		Forest; pines
Scarlet Tanager	C	C		Forest
Western Tanager		R	R	Forest
Northern Cardinal	C		C	Thickets; suburbs; forest
Rose-breasted Grosbeak	C	C		Forest
Black-headed Grosbeak		R		Forest
Blue Grosbeak	U	U		Old fields; forest edge
Indigo Bunting	C	C		Old fields; forest edge
Dickcissel	R	R	R	Fields
Rufous-sided Towhee	C	C	R	Thickets; forest edge
American Tree Sparrow		C	C	Fields; thickets
Chipping Sparrow	C	C	R	Mixed Upland
Clay-colored sparrow		R		Thickets; near coast
Field Sparrow	C	C	U	Old fields; thickets
Vesper Sparrow	R	R	R	Fields
Lark Sparrow		R	R	Thickets; near coast
Savannah Sparrow	R	C	U	Fields; SW wetlands
Grasshopper Sparrow	U	U	R	Fields
Henslow's Sparrow	R	R		Fields; wet meadows
Sharp-tailed Sparrow	U	U	R	SW wetlands
Seaside Sparrow	C	C	U	SW wetlands
Fox Sparrow		C	U	Thickets; forest edge
Song Sparrow	C	C	C	Old fields; forest edge
Lincoln's Sparrow		U	R	Thickets; fields; old fields; forest edge
Swamp Sparrow	C	C	U	FW wetlands
White-throated Sparrow	R	C	C	Thickets; forest edge
White-crowned Sparrow		U	U	Thickets; forest edge
Dark-eyed Junco	R	C	C	Forest; old fields; thickets

SPECIES	BREEDING	MIGRANT	WINTER	PREFERRED HABITAT(S)
Lapland Longspur		R	R	Fields
Snow Bunting		U	U	Fields; dunes
Bobolink	U	C		Fields
Red-winged Blackbird	C	C	C	FW, SW wetlands
Eastern Meadowlark	C	C	U	Fields; SW wetlands
Yellow-headed Blackbird		R	R	FW, SW wetlands
Rusty Blackbird		C	U	FW wetlands
Brewer's Blackbird		R		FW wetlands; farmland
Boat-tailed Grackle	C		C	SW wetlands; southern NJ
Common Grackle	C	C	U	Mixed uplands; FW wetlands
Brown-headed Cowbird	C	C	C	Mixed uplands; forest edge
Orchard Oriole	C	C		Mixed uplands; forest edge
Northern Oriole	C	C	R	Forest; forest edge
Pine Grosbeak		R	R	Forest; conifers
Purple Finch	R	C	U	Forest; forest edge
House Finch	C		C	Suburbs; forest edge; mixed uplands
Red Crossbill	R	R	R	Conifers
White-winged Crossbill		R	R	Conifers
Common Redpoll		R	R	Old fields; forest edge; birches
Pine Siskin	R	U	U	Conifers, forest
American Goldfinch	C	C	C	Old fields; forest edge
Evening Grosbeak		C	U	Forest
House Sparrow	C		C	Suburbs, farms

APPENDIX B: NEW JERSEY'S NATIVE TREES, SHRUBS, AND VINES THAT ARE BENEFICIAL TO BIRDS

Patricia Sutton

The focus of this list is native vegetation, but several alien/nonnative species (*denotes alien) that are especially beneficial to wildlife and birds have been included.

Trees

TALL (60 - 100+ feet)
MEDIUM (30 - 60 feet)
LOW (10 - 30 feet)

TREE	TALL	MED	LOW	No. of NJ Bird Species That Feed On
Red Cedar, *Juniperus virginiana*		X		30
Hemlock, *Tsuga canadensis*	X			8
PINES, *Pinus* spp.				27
Pitch Pine, *Pinus rigida*		X		28
Japanese Black Pine, *P. thunbergii**			X	—
White Pine, *P. strobus*	X			35
Scrub Pine, *P. virginiana*		X		—
Shortleaf Pine, *P. echinata*	X			—
HICKORIES, *Carya* spp.				19
Pignut Hickory, *Carya glabra*	X			—
Shagbark Hickory, *C. ovata*	X			—
Mockernut Hickory, *C. tomentosa*		X		—
Butternut, *Juglans cinerea*		X		11

TREE	TALL	MED	LOW	No. of NJ Bird Species That Feed On
Black Walnut, *J. nigra*	X			18
BIRCHES, *Betula* spp.				24
Black or Sweet Birch, *Betula lenta*		X		13+
Yellow Birch, *B. lutea*	X			13+
Gray Birch, *B. populifolia*			X	14+
American Hornbeam or Ironwood, *Carpinus caroliniana*		X		10
American Beech, *Fagus grandifolia*	X			25
OAKS, *Quercus* spp.				29
White Oak, *Quercus alba*	X			—
Post Oak, *Q. stellata*		X		—
Black Oak, *Q. velutina*	X			—
Spanish Oak, *Q. falcata*	X			—

TREE	TALL	MED	LOW	No. of NJ Bird Species That Feed On
Scrub Oak, *Q. ilicifolia*			X	—
Willow Oak, *Q. phellos*	X			—
Pin Oak, *Q. palustris*	X			—
Red Oak, *Q. rubra*	X			—
American Hackberry, *Celtis occidentalis*		X		25
American Elm, *Ulmus americana*	X			18
Red Mulberry, *Morus rubra*		X		44
White Mulberry, *M. alba**		X		44
Tulip Tree or Yellow Poplar, *Liriodendron tulipifera*	X			10
Sassafras, *Sassafras albidum*		X		23
Sweet Gum, *Liquidambar styraciflua*	X			21
SERVICEBERRIES, *Amelanchier* spp.				38
Juneberry, Shadbush, Shadblow, or Serviceberry, *Amelanchier candensis*			X	26+

TREE	TALL	MED	LOW	No. of NJ Bird Species That Feed On
Downy Serviceberry, *A. arborea*		X		19+
Wild Black Cherry, *Prunus serotina*	X			52
Pin or Fire Cherry, *P. pensylvanica*			X	42
Chokecherry, *P. virginiana*			X	43
Crab Apple, *Malus* spp.		X	X	26
HOLLIES, *Ilex* spp.				29
American Holly, *Ilex opaca*			X	13+
MAPLES, *Acer* spp.				14
Box Elder, *Acer negundo*	X			4+
Red Maple, *A. rubrum*	X			5+
Silver Maple, *A. saccharinum*	X			4+
Sugar Maple, *A. saccharum*	X			5+
Sour Gum or Black Tupelo, *Nyssa sylvatica*		X		34
Flowering Dogwood, *Cornus florida*		X		37
Persimmon, *Diospyros virginiana*		X		13

Shrubs

TALL	**(11 - 20+ feet)**
MEDIUM	**(6 - 10 feet)**
LOW	**(1 - 5 feet)**

SHRUB	TALL	MED	LOW	No. of NJ Bird Species That Feed On
Common Waxmyrtle, *Myrica cerifera*	X			25
Northern Bayberry, *M. pensylvanica*		X		25
Speckled Alder, *Alnus rugosa*		X		15
Smooth Alder, *A. serrulata*		X		15
Japanese Barberry, *Berberis thunbergii**			X	9
Common or European Barberry, *B. vulgaris**		X		7
Common Spicebush, *Lindera benzoin*	X			15
Wild Black Current, *Ribes americanum*			X	15
HAWTHORNS, *Crataegus* spp.	X			19
Beach Plum. *Prunis maritima*		X		**
Red Chokeberry, *Aronia arbutifolia*		X		12
Black Chokeberry, *A. melanocarpa*		X		7
ROSES, *Rosa* spp.				21
Pasture or Carolina Rose, *Rosa carolina*		X		—

SHRUB	TALL	MED	LOW	No. of NJ Bird Species That Feed On
Rugosa Rose, *R. rugosa**		X		—
Meadow or Smooth Rose, *R. blanda*			X	—
Virginia Rose, *R. virginiana*		X		—
Swamp Rose, *R. palustris*		X		—
BRAMBLES, *Rubus* spp.				49
Purple-flowering Raspberry, *Rubus odoratus*			X	—
Black Raspberry, *R. occidentalis*			X	—
Red Raspberry, *R. idaeus*			X	—
Alleghany Blackberry, *R. allegheniensis*		X		—
Dewberry, *R. flagellaris*			X	—
SUMACS, *Rhus* spp.				31
Winged or Dwarf Sumac, *Rhus copallina*		X		—
Smooth Sumac, *R. glabra*	X			21+
Staghorn Sumac, *R. typhina*	X			21+

SHRUB	TALL	MED	LOW	No. of NJ Bird Species That Feed On
HOLLIES, *Ilex* spp.				29
Smooth Winterberry, *Ilex laevigata*	X			7+
Inkberry, I. glabra		X		9+
Russian Olive, *Elaeagnus angustifolia**	X			26
Autumn Olive, *E. umbellata**	X			30
DOGWOODS, *Cornus* spp.				34
Alternate-leaved Dogwood, *Cornus alternifolia*	X			34
Silky Dogwood, *C. amomum*		X		18+
Red-osier Dogwood, *C. stolonifera*		X		18+
HUCKLEBERRIES, *Gaylussacia* spp.				24
Black Huckleberry, *Gaylussacia baccata*			X	12+
Dwarf Huckleberry, *G. dumosa*			X	7+
Dangleberry, *G. frondosa*		X		13+
BLUEBERRIES, *Vaccinium* spp.				36
Common Highbush Blueberry, *Vaccinium corymbosum*		X		15+
New Jersey or Coastal Highbush Blueberry, *V. caesariense*		X		—
Lowbush Blueberry, *V. angustifolium*			X	8+
HONEYSUCKLES, *Lonicera* spp.	X	X	X	21
Tartarian Honeysuckle, *Lonicera tatarica**		X		—
ELDERS, *Sambucus* spp.				46
Common Elder or Elderberry, *Sambucus canadensis*		X		33+
Snowberry, *Symphoricarpos albus*			X	8
Coralberry, *S. orbiculatus*			X	15
VIBURNUMS, *Viburnum* spp.				23
Southern Arrowwood, *Viburnum dentatum*	X			10+
Mapleleaf Viburnum, *V. acerifolium*			X	10+
Withe-rod, *V. cassinoides*		X		9+
Northern Arrowwood, *V. recognitum*	X			10+
Nannyberry, *V. lentago*	X			11+
Cranberry Viburnum, *V. trilobum*		X		6+

Vines

VINE	No. of NJ Bird Species That Feed On
Common Greenbrier, *Smilax rotundifolia*	20
Cat Greenbrier, *S. glauca*	19
Poison Ivy, *Rhus radicans*	33
American Bittersweet, *Celastrus scandens*	15
Porcelainberry or Asiatic Ampelopsis, *Ampelopsis brevipedunculata**	13
Virginia Creeper, *Parthenocissus quinquefolia*	35
GRAPES, *Vitis* spp.	51
Fox Grape, *Vitis labrusca*	—
Summer Grape, *V. aestivalis*	—
Riverbank Grape, *V. riparia*	—
Frost Grape, *V. vulpina*	—
HONEYSUCKLES, *Lonicera* spp.	21
Japanese Honeysuckle, *Lonicera japonica**	—
Trumpet Creeper, *Campsis radicans*	2

Key

*	alien/nonnative
**	highly important secondary food source (insects: warblers)
—	see previously noted number of users for family of plant
+	indicates additional species known to feed on family of plant

FLIGHT PATH

References:

Anderson, Karl. 1979. *A Checklist of the Plants of New Jersey.* Mount Holly, NJ.: Rancocas Nature Center (NJ Audubon). 51 pp.

DeGraff, R. M. and Whitman, G. M. 1979. *Trees, Shrubs, and Vines for Attracting Birds.* Amherst, MA.: University of Massachusetts Press. 194 pp.

Martin, Alexander C., Zim, Herbert S. and Nelson, Arnold L. 1951. *American Wildlife and Plants: A Guide to Wildlife Food Habits.* New York, NY.: Dover Publications, Inc. 500 pp.

Petrides, George. 1958. *A Field Guide to Trees and Shrubs.* Peterson Field Guide Series #11. Boston, MA.: Houghton Mifflin Co. 1958. 428 pp.

Sutton, Patricia (Naturalist, Cape May Bird Observatory). Personal field notes. Cape May Bird Observatory (NJ Audubon), P.O. Box 3, Cape May Point, NJ 08212.

MERLIN
FRANK SCHLEICHER

Front Cover Photos (clockwise from the top)—The Lady and Snowy Owl (FRANK SCHLEICHER); Hudsonian Godwit (FRANK SCHLEICHER); Black-legged Kittiwake (RICHARD L. DITCH); Tree Swallow (FRANK SCHLEICHER); Sharp-shinned Hawk (PETE DUNNE).

New Jersey Audubon's Five Centers

Lorrimer Sanctuary

Scherman/Hoffman Sanctuaries

Owl Haven Nature Center

Rancocas Nature Center

Cape May Bird Observatory

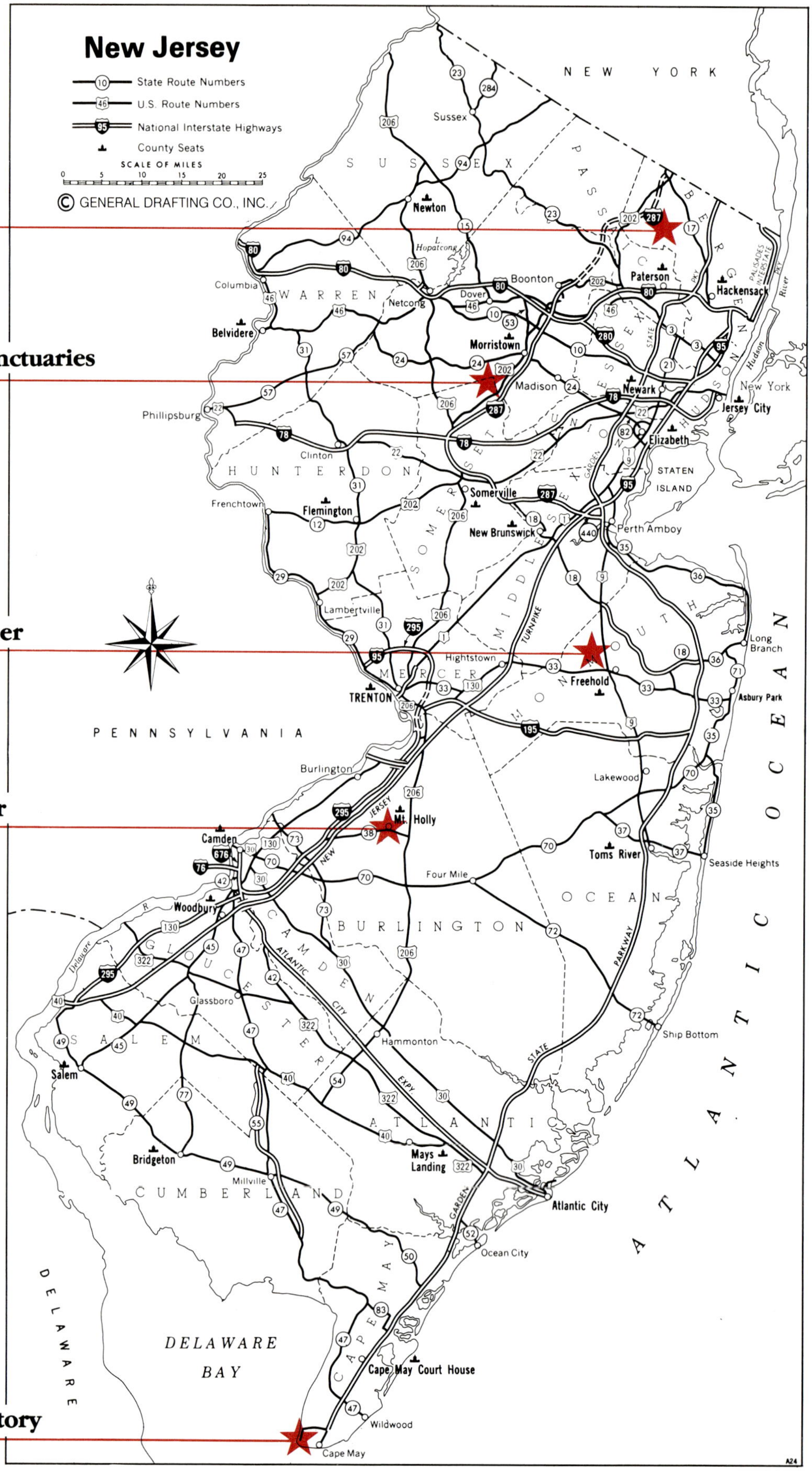